SHARING IT ALL

The Rewards and
Struggles of
Two-Career Families

SHARING IT ALL

The Rewards and Struggles of Two-Career Families

Lucia A. Gilbert, Ph.D.

PLENUM PRESS • NEW YORK AND LONDON

Library of Congress Cataloging in Publication Data

Gilbert, Lucia Albino.
Sharing it all: the rewards and struggles of two-career families / Lucia A. Gilbert.
p. cm.
Bibliography: p.
Includes indexes.
ISBN 0-306-42961-6
1. Dual-career families—United States. I. Title.
HQ536.G533 1988 88-19478

306.8'7—dc19

Plenum Press is a Division of
Plenum Publishing Corporation
233 Spring Street, New York, N.Y. 10013

Printed in the United States of America

To my husband, Jack, my daughter, Melissa,
my mother and father, Carmelena and William,
and my mother-in-law, Nellie

PREFACE

I realize more and more that to some degree psychologists study their own lives. My first studies in the mid-1970s looked at the conflicts that emerge in dual-career couples as the partners combine their various roles. Ideas for the early studies initially came from conversations with clients, friends, and people I met traveling. Soon after the topic of dual careers came up, discussions of guilt and frustration followed. The participants in my first studies turned out to be predominantly women. Men expressed little interest in participating. A common response was, "Talk to my wife."

Ostensibly, husbands saw their partners, but not themselves, as dealing with conflicting roles. Although I presumed this to be somewhat true, I knew from observing my own husband that my having a career had an immeasurable impact on him. Were men denying something? Were women overly sensitive? Discussions with my husband, which helped me to get a better grasp on what male partners might be feeling, planted the seed for my subsequent book on men in dual-career families, published in 1985.[1] In gathering material for that book, I was struck by the contrast with men's earlier disinterest. Husbands appeared eager to talk about their lives. Their greatest concern—what happens to the children?—became the focus of my last set of studies.

Why this book? "Sharing it all" symbolizes the essence of a two-career marriage. Yet succeeding at this kind of a marital relationship goes against convention and represents dramatic social change. So far, observers of the dual-career family have overemphasized its impact on women and the costs to them. Little attention has been given to the subtle revolution in men's lives. My research over the past 15 years with the women, men, and children in two-career families may provide the reader with a broader understanding of this phenomenon. In this book I explore *both* the husband's *and* the wife's experience and focus on how the attitudes and structures of society affect each partner in this revolutionary type of marriage.

The book details the unique perspectives women and men bring to the dual-career marriage and how these perspectives can contribute to possible struggles within the marriage. The two chapters in Part I, "Women's Slowly Changing Reality" and "Careers and Caring: Can Men Manage Both?", introduce the premises that underlie the socialization process. Part II, "Living as a Two-Career Family," describes the two-career family from three perspectives, each with its own chapter—the couple's marital relationship, children, and the two careers. Themes introduced in the earlier chapters get attention here as well.

"How to More Than Manage"—the topic of the final part—focuses on strategies for sharing it all. Certain things go with the "dual-career turf"; knowing what to expect provides a sense of well-being and control. Because the success of the two-career family lifestyle requires social change, it brings up conflicts not only within individual women and men but also between the two-career couple and the world in which they live. Couples who want to "share it all" feel resistance from within and without to acting in accord with their agreed-upon principles of equity or sharing. They function in a larger society in which career obligations and support systems for child care remain at odds with the couple's private principles of equity. Chapter 7, "Stress and Coping in Two-Career Families," discusses how couples can prepare themselves for and effectively cope with possible and inevitable stresses. At times, professional help can be extremely useful. Deciding on such help and the effect it might have are discussed in Chapter 8.

Despite the struggles, more and more couples wish to succeed as a two-career family. The rewards, which include freedom from overly restrictive sex roles, make the struggles worth it. But, as the last chapter makes clear, ultimately the viability of the two-career family rests on couples being willing and able to share it all.

Lucia A. Gilbert

ACKNOWLEDGMENTS

I doubt this book would exist were it not for my husband Jack and my daughter Melissa. In many ways I grew up with them and they grew up with me. Jack and I knowingly entered into a dual-career marriage. We knew it was possible but we had to learn to make it succeed. Melissa's birth changed our lives and made us deeply aware of how much our personal ambitions included close connections with a child. I especially appreciate Jack's and Melissa's belief and faith in me—as a wife, mother, and psychologist. There would be no story to tell without them or, for that matter, without my mother and father, Carmelena and William Albino.

Reared at a very different time and in a different cultural environment, my parents never faltered in their support of a lifestyle dramatically different from their own. My mother-in-law, Nellie Gilbert, provided generous emotional support over the years and always took a positive stance toward our dual-career struggles.

Particular gratitude goes to all those people who let me in on their lives. In the book I draw on the experiences of family, friends, students, colleagues, patients, and participants in research studies and workshops. An academic leave from my position at the University of Texas at Austin last winter proved to be especially beneficial. I met Sherwin Davidson and her husband, Michael Toth, a dual-career couple who themselves have written on this topic. Our many lively discussions helped convince me that I should write this book. Numerous conversations with many friends and colleagues, especially Connie Deutsch, Vicki Rachlin, Michael Newman, and Murray Scher, clarified my thoughts and pushed my thinking further. Earl Koile, my long-time mentor, kept me on track and let me know at just the right moments that what I was doing was good. Finally, Mary Dewar, a special friend and accomplished historian, must be thanked. Among other things, she knew exactly how to raise my spirits: Invite us to a sumptuous Chinese dinner prepared by her husband Michael and then during dinner express confidence in how splendid my book would be.

Linda Greenspan Regan of Plenum Publishing Corporation also made this book possible. She contacted me about doing a book on professional women. I said I would be interested in the topic of professional women *and* professional men—two-career families—but not now, perhaps later. I just did not have the time. She persisted. I changed my mind after nearly a year of telephone calls and letters back and forth between New York and Texas.

Very special appreciation goes to Victoria Cherney, Assistant to the Editor. She was relentless in encouraging me to step away from my teacher–researcher style of writing to a style that communicated more directly with the reader. Her unwavering interest in the material and her insightful critiques of chapters proved invaluable. I feel fortunate to have had the experience of working with her.

L.A.G.

CONTENTS

CHAPTER 1

A PERSONAL JOURNEY

The time is August 1974, the last day of our year of "living together, apart." My husband and I want to remember as purely as possible what this year has been like for each of us—the joys and traumas, the ups and downs. Rather than talk about it or tape-record reflections, with pen and paper in hand we go into separate rooms to reflect and write.

Our separate stories, written that day over 14 years ago, appear in this first chapter because they speak so candidly of the feelings and realities of an actual working relationship—our own relationship. We entered into our own variation of the dual-career lifestyle for reasons that, although they were different, focused on a common goal. Not unlike most other dual-career couples, we wanted our relationship to survive and took steps that we hoped would ensure that eventuality. Another reason for communicating this experience with others is to let them know that a commuting lifestyle, even for a relatively short period of time and despite the obvious difficulties, can be inherently beneficial and rewarding for both husband and wife, mother and father, and even for the child.

Finally, and perhaps most importantly, I want to depict how my husband shared with me in addressing my needs for independence and making it on my own. The significance of this cannot be emphasized too much and gets discussed at several later points in the book. Supporting another person's need for independence requires a good deal of security and confidence in oneself. It's difficult if not impossible to do when the person who is asked to provide the support feels scared, threatened, angry, or not sufficiently supported himself or herself.

My Account as of August 1974

I suppose one could call our recent lifestyle one of remote parenthood. I live in one city and work at the major university there. My husband is a professor at a different university in a different city. Our daughter lived with her dad until she was five months old and then with me until now (seven months old). The past year has been exciting and rewarding in many, many ways. It also has been emotionally trying and lonely.

The beginning was sudden. I was about to complete my Ph.D. in psychology at the same university at which my husband had been employed for eight years and was now tenured. One day I decided that my career opportunities could not be limited to that *one* city. I knew I had a good chance of obtaining teaching positions elsewhere and that the opportunities where we now lived were extremely limited. My husband supported my decision, and I proceeded to apply for positions outside our present locale with the proviso that each of the universities I selected also have a good department in my husband's field, chemistry. I found just the job I wanted—combining teaching, research, and psychotherapy. And they wanted me. Well, it was one thing to think that one wanted to make it on one's own and another to take the step. We decided to live apart for one year, alternating commutes every other weekend, with the hope that he could find a position the following year at the same university at which I had accepted a position.

This was April and the appointment was for September. I continued to work furiously on my dissertation and to feel very excited about the future. In June I discovered that I was pregnant—a shock to both Jack and me. I had never before and have never since experienced such a deep depression. Yes, I had wanted to have a child some day. In fact we had tried unsuccessfully to have a child during the two years I was finishing up my doctoral degree. Subsequently, we had decided to postpone things for a few years and had gone back to using birth control—a stratagem that Nature had now effectively countered. But now was not the time to have a child. Psychologically I was prepared to go off on my own. In fact, the thought of never having a child had recently crossed my mind for the first time.

My husband, too, was deeply depressed. Abortion, an obvious solution, did not fit for us. We knew all along that we desired a child. Moreover, the disappointments of the previous two years sensitized us to the possible risk of not going ahead now. The question slowly shifted from whether to have the child to how we could accommodate a child in our life of separation. We decided to be flexible and to go ahead with our career decisions as planned. I would take a leave of absence for the month of January, the month when the baby was due. My parents would care for the child for two weeks and my husband's mother for two. This would take the child up to the age of two months, an age suitable for placement in the small day-care center we had located. I would then commute back every ten days for a four-day weekend.

Much to our surprise, our parents showed enthusiasm and support for our plans. Our friends, however, seemed more skeptical, and several expressed concern about a child being without its mother for over a week at a time. I admit that I too was worried, but I sensed we were making the right decision. I knew that I had to be my own person professionally and that I very much wanted to have a child. I also knew that choosing one experience over the other would leave me resentful. My mother always told me, "Where there is a will, there is a way," an adage that fit this situation perfectly. The decision made, we proceeded.

My new colleagues were rather surprised to discover that I was pregnant. I often wondered whether they thought I was a bit weird—particularly when they learned that Jack was to be the "mother." The pregnancy in no way interfered with my work. On the contrary, I think it had many positive effects—especially as a modeling experience for female undergraduate and graduate students, and perhaps even for my male colleagues and students.

Time flew. The birth was arduous, beautiful, and natural, with the father participating as well. (My husband and I had separately taken Lamaze classes.) My daughter is a beautiful child. From the moment of birth she was alert, happy, and delightful. But leaving her was incredibly difficult. I had no idea that such a strong attachment could be formed in such a short time. My husband forced me to set a day to go and to make plane reservations. He stated emphatically that I would leave on that day regardless. Three weeks after her birth I left for the first time. I cried all the way to the airport and much of the four hours on the plane. On returning to work the next day, I was greeted with tremendous warmth by graduate students and the younger members of the faculty and staff.

In reading over what I wrote so long ago, I find that carrying out our plans, once they were made, sounds easy. It was not. The easy parts were related to my professional life. I worked day and night loving every minute of it. The hard parts were the loneliness interspersed with every other weekend together with Jack (and later Melissa) followed once again by separation. I had anticipated being alone but was not prepared for the loneliness in terms of interpersonal relationships. My new colleagues at the university, although warm and friendly, showed little interest in interactions beyond the formal professional aspects. I realized that a woman living apart from her family seemed a bit unusual, and perhaps they saw me as somewhat of a social misfit. As things turned out, my two closest friends were both male colleagues, one of whom was in a dual-career family and had two small children. The other was single. They came to be the people with whom I could share both my professional and my personal self, and their support and friendship really helped sustain me through a difficult period.

I enjoyed time for myself. I missed, however, the physical and emotional closeness of my husband. Being apart made me realize how much we communicated, how much we enjoyed one another intellectually, and also how emotionally dependent I was on him. In our relationship I am the effervescent one, he the stable, logical one. I missed his stability. The commuting tended to make our relationship extremely intense. We had to squeeze 11½ days' worth of communicating into 2½ days. Sundays came to be hated days, a time for immediate readjustment, ready or not. We knew it would be easier if I commuted less frequently, but our time together seemed invaluable, particularly at a time when so much was happening in our lives. First, I was adjusting to my first professional position. Second, we were expecting and then had our first child. Third, our future life lacked certainty.

What I most gained from the experience is an appreciation for what it means to do "what feels right." Having had this opportunity, I now know that I want a life apart from my husband and child, and that I also need to live with them as much as I need this other part of me.

My Husband's Account as of August 1974

I have long held the belief that we, as a society, waste a significant component of our total intellectual potential by hewing to the conventional view that women first and foremost are housekeepers and mothers, only secondarily accepting the nontraditional possibility of their potential as professionals. What or who planted the seed of this belief in me is not clear, but the belief has remained strong even in the face of innumerable arguments with many male acquaintances who, for whatever reasons, prefer the more traditional functions of women in our culture.

Consistent with this belief, I encouraged my wife in her efforts to pursue a professional career, doing so even before we were married. A major question, however, was just what kind of career would be the most fulfilling for her. It was clear to both of us that she was not going to be happy with a profession built around chemistry—a field in which we both had postgraduate training—a fact that she quickly discovered upon teaching that subject for two years. A Ph.D. degree in the field of counseling psychology was the entirely unexpected upshot of her aborted teaching career in chemistry.

What an experience in personal growth the graduate program leading to that degree turned out to be! What I perceived in her at the time she was just completing her degree was a person who had done exceedingly well in her formal graduate academic work, at least to the extent that course grades and the respect of colleagues measure such success, and who also had achieved unusual skills for effective individual and group therapy. In marked contrast to her approach to chemistry, she appeared supremely confident of her abilities in the field of counseling psychology. In my eyes, not only her professional but also her personal growth had been nothing short of phenomenal during the four years prior to the awarding of her Ph.D. I felt that I was now married to a woman who was both intellectually and emotionally an independent person. Nevertheless, a primary factor in my own thinking with respect to the advisability of undertaking a temporary commuting marriage was my impression that she possibly did not fully appreciate the full scope of her maturation and independence; the act of living apart, I hoped, would bring the lesson home more fully to both of us.

Suffice it to say that I was very unsure about the soundness of our decision with all its possible ramifications, and my uncertainty increased greatly when we learned that my wife was pregnant with our first child. Neither one of us was at all confident that we would be successful at handling the loneliness and unrest that were bound to develop by living apart, particularly since our marriage, I think it fair to say, had developed into a very intimate, shared type of experience. Furthermore, it was she who was being forced to plunge into an entirely new environment, both socially and professionally, and would have to deal with the resulting pressures as well as with her pregnancy mainly on her own. My situation would be changed, to be sure, but not nearly so dramatically as hers. The possibility seemed great that she would not be able to manage all of the new stresses, but, although I think we both recognized that our experiment might be a spectacular bust, we did not verbalize our doubts.

More than once we came close to "throwing in the towel" with respect to continuing to live apart. The strain for her of going through a first pregnancy with no husband with whom to share on a daily basis the joys and anxieties of this most fundamental of human experiences was an early and ongoing problem. Long-distance phone calls are no substitute for face-to-face interactions, and I was extremely frustrated by having to offer what moral support I could by way of this incredibly sterile device. However, probably the most difficult soul-searching of all occurred when it came time for her to leave our three-week-old baby with me and resume her career far from home. Once again, it was she and not I who had to bear by far the greater burden emotionally. The decision whether or not to return now seemed entirely hers. I guess it is a measure of her own recognition of her strengths, both personal and professional, that she did again undertake the commuting style of living that was now so painfully familiar to both of us.

Looking in retrospect at our total experience, I'm glad that we attempted it and that I was able to support her needs. We both learned a lot about how

> much we give to one another and how this sustains our ability to continue on our seemingly separate professional tracks. Living through the emotional swings of the past year gave us a new strength and somehow instilled a feeling that we can survive almost anything, although the road ahead promises to be rocky. We will continue to struggle with our ambitions and finding positions to our liking in the same locale. Overall, though, one could not have hoped for a more positive outcome.

Coda

These vignettes reflect one of many possible outcomes. They describe the events in the life of one two-career family. Yet the bridge between the unique and the universal seems wide. I know from experience that other women share my ambitions, fears, and struggles. I am not all that unique. My husband reports similar although much less frequent experiences. Couples embarking on a two-career marriage or encountering some difficulties often ask my husband and me the "secret of our success." They expect some easy answer and show surprise that we struggled and still struggle at times. We look so happy. What we say typically strikes familiar chords in their own lives and gives validity to the pervasiveness of our situation. I suppose in some ways this book spells out in still greater detail my response to that question.

Many complex factors influence the extent to which a shared and mutually supportive relationship gets achieved. I will be addressing in the following chapters the questions of what fuels the pursuit of this lifestyle, what realities lie in store for partners, and which ways they best manage both careers and marriage successfully.

PART I

THE SETTING: WOMEN AND MEN

Wife and husband each bring a unique perspective to the two-career marriage. The rewards and struggles that ensue must be considered from both standpoints.

CHAPTER 2

WOMEN'S SLOWLY CHANGING REALITY

> While the large majority of Harvard men have enjoyed professional success, bringing home even larger incomes, the career paths of many Radcliffe women have been broken, interrupted by marriage, children, and what some of them argue is a hostile, male-dominated world.[1]

Combining a professional career with family life poses difficulties for women. This may come as a surprise to some. After all, feminism supposedly won its battles in the 1960s and 1970s. There should be no problem—

egalitarianism surely prevails in two-career relationships and everything gets shared.

But, as the introductory quote clearly indicates, fundamental change comes slowly. The lives of men and women remain strikingly different, even today. Women's experiences growing up, while quite different from those of women historically, still prepare them for a world in which women and men act differently and in which women preside over family matters. Having a career and a family still represents "trying to have it all" for women.

The results from a recent study illustrate this slowly changing reality for many women.[2] The researchers followed the lives of high school valedictorians who graduated in 1981. All the young women and men in their sample went on to college. The women outperformed their male counterparts academically all the way through college, but considerably fewer women than men fulfilled their original career goals. Plans to become physicians, scientists, and trial lawyers changed to include careers in teaching, nursing, and social service. Moreover, once in college, many fewer women than men considered themselves intellectually more talented than their peers. Thus, intellectually gifted young men and women who held similar aspirations as seniors in high school differed considerably by the time they graduated from college.

These differences in the actual lives of women and men, although in themselves neither necessarily good nor bad, do not come about accidentally. We would be naïve to think that women and men make life choices solely as *individuals*. Situational pressures and social norms and practices help mold what appear to be individual decisions about career and family and whether and how they combine. Women and men do not live in a vacuum. Rather, the two sexes are brought up to think of themselves and each other as opposite sexes with nonoverlapping abilities and life goals. Their experiences in childhood and in various educational environments reinforce these views. Women, for example, receive less encouragement from professors and encounter relatively few female professors as role models for future professional pursuits. Also, women, not men, perceive themselves as having to contend with decisions about how to combine family with a career. No one totally escapes the tentacles of "opposite sex socialization"—not even the women (and men) who choose dual-career marriages.

Concepts of Women and Men as Opposite Sexes

It still surprises me when I see blue and pink bows Scotch-taped to the heads of newborn babies displayed in hospital nurseries. This example, trivial as it may be, reflects a larger and very important reality. Biological sex makes a difference in our society and in many ways gets emphasized from the moment of birth. Parents raise girls to be kind, loving, and nurturing. Parents and teachers alike expect girls to be well-behaved, neat, and orderly, and they comment when girls act otherwise. Their future is pretty well mapped out—girls eventually marry and have children. My daughter was first proposed to when she was five. The checkout man at the grocery store asked her to marry him and promised he would wait for her to grow up. She has probably received an average of five proposals a year since that time.

The cartoon strip "Cathy" reminds us daily that many women believe they need men desperately.[3] Cathy's thoughts center on filling her "inner space"—either with food or through a relationship with a man that brings eventual motherhood. She feels incomplete without a man and poignantly plays out in an exaggerated way common views of women's dependency on men.

Boys, on the other hand, are socialized to develop freedom of spirit, independence, and self-directedness. In contrast to their expectations for girls, parents and teachers expect boys to break the rules, be indifferent about their attire, and let others pick up after them. Their future is also mapped out—to be successful in careers and have a family. How cute boys are is far less important than how talented or smart they are. Boys rarely receive proposals of marriage from checkout women at the grocery store.[4]

But are the sexes truly "opposite?" In reality, there are few, if any, ways in which women and men are really opposite. Different, yes; opposite, no. For instance, as babies, males and females cry about the same, but as adults, women generally cry more than men. The sexes learn to express emotions in somewhat different ways, but both sexes possess the capacity to cry. Weight provides another good example. Women typically diet more than men, although a comparable

cathy
AHAH!
WHAT ARE YOU READING, CATHY?
"WHY MEN ARE THE WAY THEY ARE," BY DR. WARREN FARRELL.
OH YEAH?
I ALSO HAVE "SUCCESSFUL WOMEN, ANGRY MEN"....
"WOMEN WHO LOVE TOO MUCH"... "MEN WHO CAN'T LOVE"....
"SMART WOMEN, FOOLISH CHOICES"... "MEN WHO HATE WOMEN AND THE WOMEN WHO LOVE THEM"....
IT'S A WHOLE NEW LINE OF BOOKS THAT HELPS US STUDY THE NEEDS AND MOTIVATIONS OF THE OPPOSITE SEX SO WE CAN WORK TO CREATE MORE MUTUALLY FULFILLING RELATIONSHIPS.... WHAT ARE **YOU** READING, IRVING?
"HOT ROD WORLD."
I THOUGHT YOU WANTED TO UNDERSTAND ME.
I ALREADY KNOW TOO MUCH.

proportion of women and men fall into the "overweight" category. This difference in their dieting behavior does not come about because women occupy the fat end of the weight continuum and men the thin end. Rather, for reasons connected with socialization and body image, women worry more about the extra pounds. Generally speaking, men feel less pressure to achieve a svelte look.

Nonetheless, we unwittingly use the term "opposite sex" and act *as if* women and men fundamentally differ. We learn from an early age that the attributes that describe girls should not describe boys, and vice versa. Incorrect assumptions based on stereotypes lead us to believe that women are dependent and emotional, men are not and should not be; men are independent and unfeeling, women are not and should not be. The stereotypic list goes on and on. Women are intuitive, weak, and virtuous; men are rational, strong, and capable of deceit. Women take care of children and the home; men take care of women and children and the world outside the home.

The expression "opposite sex" implies even more—that women and men presumably act in opposition to one another. It suggests an underlying antagonism or basic conflict—the pitting of one sex against the other—men against women, women against men, the proverbial battle of the sexes. All too often we hear comments intimating that women "catch men," "hook men," and devote their lives to trying to change men's erring and irresponsible ways. Men, according to the lore, fear "losing their independence," "getting tied down," and being "hooked" by women. As the "Cathy'' cartoon reproduced earlier illustrates, because women supposedly need men more than men need women, they must continually strive to understand men in order to hold on to them.

This kind of thinking profoundly affects individuals in dual-career families. They too grew up in a society that views women and men as fundamentally different, if not opposite, and they too internalized traditional notions of what it means to be a woman and a man. The kind of marriage they aspire to, based on an egalitarian relationship between husband and wife, stands in marked contrast to convention and to deeply ingrained beliefs about how men and women interrelate. Later chapters identify areas of predictable conflict and ways they can be resolved.

Female Dependency and the Overvaluation of Love

> I wanted to write about a girl in love—not just with a man, but with the world.[5]

This statement about a 17-year-old heroine in the novel *Rich in Love* alludes to the limitations society sometimes places on women. The author wanted to present an adolescent who goes beyond the stereotype, one capable of mastering the intrinsically healthy conflict between her desire for security and dependence and her thirst for novelty, adventure, and independence.

To a large degree women have been conditioned to focus their lives around men. Their efforts to act autonomously and to broaden their interests and activities beyond the sphere of the family were met with great skepticism historically—independent achievement supposedly ran counter to "women's nature." They did not exist apart from men. This attitude did not reflect women's reality, however, but rather represented the patriarchal ideal of womanhood. It benefited men to perceive women as primarily longing to love a man and be loved by him, to admire him and serve him, and even to pattern herself after him.[6] Women's obvious dependency on men was not innate, however. It was learned. Women admired and served men not because that's their nature, but because tradition and male power shaped women's choices and behaviors.

Even today, views of a successful woman require that she attach to or depend on others—in particular, that she affiliate with and attach to a man. Many people wonder what good a career is to a woman if she has no one to come home to. Women without husbands evoke pity. Girls still learn from an early age that their main role is to attract men as life partners; containing or limiting their own achievement may become necessary if relationships with men become endangered. A competent, worldly, and ambitious woman could invite the kiss of death because such women are less desirable to men as life partners. Men ostensibly prefer women who need them.

A case in point is the recent Harvard–Yale study on the marriage prospects of women over 30.[7] Although the motivations of the researchers remain unclear, no mention was made about attempts to gather parallel data on the marriage prospects of men over 30. The study, released in 1986, asserted, incorrectly as we will see later, that women who postpone marriage beyond their 20s may never find a mate. The study gave white, college-educated 30-year-old women just a 20% chance of marrying. Women at 35 were given a mere 5% chance; at 40, the odds dropped to 1.3%. For black women over 30, the odds were considered even dimmer.

These dismal prospects caused considerable hysteria among many women I know and even resulted in some women reconsidering their career decisions. One of my students, a single woman in her 30s, turned down a demanding position for one that would leave her more time for a social life. She heard this study reported on television and decided to maximize her availability. A 36-year-old colleague confided that the day she heard the report on the radio, she turned down an attractive offer that would have taken her out of a metropolitan area with a large professional population. She wondered later whether the statistics, which she intuitively knew could not be correct, had unconsciously influenced her decision: "I think I am beyond that. But the idea of never finding a mate really scares me and makes me angry at the same time. Wanting a relationship is natural. But the spotlight stays on women like me. If we never marry, people say it's our own fault. 'You made the wrong decision. Tough luck, lady.' It also galls me that anyone would conduct such a study. What's the point of it—to scare women into traditional roles?"

A later study by the Census Bureau differed with the Harvard–Yale predictions.[8] Its calculations indicated that single 30-year-old women have a 58–66% chance of getting married. And at 40, the estimated chance was 17–23%. This later set of projections may or may not be accurate. Nevertheless, the societal message to women is loud and clear. Unlike men, women may not be able to have a career *and* a family. Developing a career, which may require postponing marriage, makes women *less* desirable as partners but renders men *more* desirable. Moreover, men can marry after 30 without worry. Their age does not necessarily jeopardize prospects of marrying and their biological clock with regard to childbearing ticks much longer.

The emotional upset caused by reports of the Harvard–Yale study caught some women by surprise. Was this sufficient reason to contemplate or reconsider career decisions? Did they have to give up their careers to get a man? Should they have married earlier and run the risk of not finishing school or establishing careers? Were they neurotically dependent on a man? Were they no different from Colette Downing, author of *The Cinderella Syndrome,* who nearly gave up her professional ambitions as a writer once a man slipped a gold ring on her left finger?[9] She derived the title of her book from the prince's putting a glass slipper on Cinderella's left foot.

Such self-doubts arise as a result of female *and* male socialization. Women want a different kind of relationship with men and in the world but they feel resistance from both men and society. On the one hand, many women know careers and relationships fit for them. On the other hand, they wonder whether traditional views reflect some basic kernel of truth about women and men. The popular press and the general culture feed these doubts. We can see evidence of this in the "Cathy" cartoon, in which Cathy enumerates the latest books that help women "study the needs and motivations of the *opposite* [emphasis added] sex so we can work to create more mutually fulfilling relationships."

Women often fall into traditional patterns with men without even realizing it. It is not unusual for women to enter into relationships with fairly "enlightened" men only to find that they have slipped into picking up dirty socks and worrying about meals while "he is out playing golf." Still today, women may consciously or unconsciously relinquish or hide their own strength and protect men from the knowledge that they can take care of themselves. Some fear, perhaps realistically, that women are less desired by men when they are economically independent and personally assertive. Reports like the Harvard–Yale study confirm their apprehensions.

Consistent with female socialization, then, some women may assume a dependent stance in relationships with men as a way of having a career and family. This underfunctioning, which primarily serves to protect men from women's ability to act autonomously, also serves to protect women.[10] That is, women may believe they need to subordinate themselves to men in physical and/or intellectual contexts regardless of their actual capabilities. By so doing, they can assure themselves a safe place with men. Because underfunctioning as a strategy has its roots in traditional aspects of female and male socialization, certain marital patterns may be unintentionally developed between partners without either partner recognizing what actually motivates his or her behavior. In one situation, for example, the wife handled all the finances, including investments and monthly bills, and routinely consulted with her husband on these matters. However, consistent with traditional views, they both saw him, not her, making the decisions about money.

The notion that women must strengthen men by relinquishing or hiding their own strength is not new. We see poignant examples in fairy tales, movies, and real life. A vivid illustration of female underfunctioning occurs in the short story "Barcelona."[11]

> The story describes Persis Fox, a "fairly successful illustrator" who is beginning to be sought after by New York publishers but sees herself as cowardly and fearful. Her husband Thad, in contrast, is a self-assured, self-directed, confident man who teaches at Harvard. The couple are on vacation in Barcelona and on their way to a remote restaurant when a thief suddenly snatches Persis's purse. Thad, who has been quite unattentive to Persis, suddenly springs into action and runs after the thief despite the darkness, danger, and unfamiliar terrain. The purse is recovered on the cobbles and the couple return to the restaurant.
>
> Thad asks, "Aren't you going to check it? See what's still there?"
>
> . . . "Oh good, my passport's still here," she tells Thad.
>
> "That's great." He is genuinely pleased with himself—and why should he not be, having behaved

> with such courage? Then he frowns. "He got all your money?"
>
> "Well, no, actually there wasn't any money. I keep it in my pocket. Always when I go to New York, that's what I do."
>
> Why does Thad look so confused just then? A confusion of emotions is spread across his fair, lined face. He is disappointed, somehow? Upset that he ran after a thief who had stolen a bag containing so little? Upset that Persis, who now goes down to New York on publishing business by herself, has tricks for self-preservation?
>
> "But you said your passport's still there?"
>
> "Oh yes, of course," Persis babbles. "That would have been terrible. We could have spent days in offices."
>
> Gratified, sipping at his wine, Thad says, "I wonder why he didn't take it, actually." Persis does not say, "Because it's hidden inside my address book"—although quite possibly that was the case. Instead, she says what is also surely true: "Because you scared him. The last thing he expected was someone running after him, and that *whistle*."
>
> Thad smiles and his face settles into a familiar expression: that of a generally secure, intelligent man, a lucky person, for whom things happen more or less as he would expect them to.[12]

Thus, Persis protects Thad from the knowledge that his motivations for protecting her against worldly dangers are unfounded and that, in essence, she knows how to take care of herself. She also realizes that Thad's reasons for protecting her relate more to his needs than to her needs: In the story, Persis thinks, "He is not doing this [the chase] for her; it is something between men." They both collude to maintain a system that shelters the male ego from the threat of female competence and that continues an illusion of female dependency. Thad does not want to see Persis's recently developed independence and competence in dealing with the external world, and she makes little effort to show it to him.

Persis Fox's dilemma and insight parallel the experiences of many women in dual-career marriages. Some women may even recognize how they take a dependent stance with men but consider it necessary and useful. The tradeoff or cost, of course, entails taking a one-down position vis-à-vis the man. One woman reported that her husband complained about not feeling needed by her. She responded by centering her life around him more, which helped matters immeasurably from his perspective. In the process, however, she compromised egalitarian goals and even developed symptoms of depression.

Concepts of Achievement

> I realized all along that I had little sense of direction despite my obvious talent. I also knew I wanted to marry and did so right after college. Now seven years later I feel isolated, angry at myself for not pursuing my career as an artist, and for no longer being in love. [Patsy, a woman in her early 30s who had just returned to school]

The notion of female dependency also influences views on how best to prepare women for their future life roles. Traditionally, the choice seemed to be, "Shall we socialize women to be competent or to be feminine?"[13] The goals of femininity and competence were presented as mutually exclusive. It was implicitly assumed that women reared to be feminine need and want men to take care of them. In response, they center their life goals around men and children and take care of others for self-fulfillment. Competence, on the other hand, requires independence, self-sufficiency, self-reliance, and self-direction. Women reared to be competent would possess the means sufficient for the necessities of life—they could take care of themselves. But according to traditional thinking, would they still be feminine? Would they be "real" women?

Women today still encounter these narrow attitudes. They worry about losing some key aspect of their womanhood if they succeed in careers. I vividly recall an ad I saw not too long ago. A woman dressed in a business suit said, "I was never so insulted in all my life. He said he loved me for my brains." This ad, clearly demeaning to women yet designed to appeal to them, provides a clear example. The message to women is that men should not find you attractive for reasons other than your beauty, and certainly not for reasons associated with intellectual competence.

Patsy's situation provides another illustration of messages to women that in effect say women cannot be "who they are" and have careers too. The earlier discussion on female dependency pointed out how women traditionally learn from an early age that men are superior and that women need men to take care of them. By underfunctioning, women hide their competence from themselves and from men. Doing otherwise would jeopardize this security and force women to make it on their own—which in effect means without a man and with a good deal less financial security. Moreover, what women witness daily reinforces the view that they can't do as well as men professionally. Men still hold the most prestigious positions and control all levels of government in most countries.

The notion that the sexes are "opposite" causes many women (and men) to regard "womanliness" or femininity and competence as incompatible. Simply put, competence, an attribute of men who do things out in the world, does not describe women, who focus their work within the family. This kind of dualistic thinking obfuscates the obvious: Competence knows no gender or setting—it belongs to neither men nor women and describes members of both sexes. Thus, competence and femininity can and do coexist. It's all in the eye of the beholder. Women restricted to traditional female roles, for example, act with independence and self-direction in caring for their families. They tend to the home, work in positions outside the home, and rear children, often single-handedly. But if competence belongs only to men and male roles, women, regardless of their accomplishments and abilities, cannot be considered competent in the patriarchal sense of the word. They can be competent mothers and wives, but that does not count as true competence.

This seeming either/or situation can pose problems for women in dual-career families. Career-oriented women and men do not differ in competence. But only women have to integrate independence and self-direction, which are now more acceptable for women, with traditional notions of femininity. Not surprisingly, then, women show greater sensitivity than men to what professional success might cost them in other ways. Still today, women with families who succeed in moving up career ladders get special press. They are the exception, not the rule. This reality affects women's views of themselves and their possible life goals. My students tell me that I give them hope. Knowing that I successfully combine work and family makes it easier for them to imagine it for themselves. They are hungry for role models.

Women's sensitivity often exists at an unconscious level. Career goals often remain less defined and more associated with marriage and family than a woman realizes. Take the case of Patsy. In graduate school she fell madly in love with John, an ambitious and competitive man in the same field as she. Patsy was the more talented artist of the two, and this bothered John a lot. He mentioned how much he envied her natural ability and she reassured him that he was as good. She assumed their mutual love would enhance the development of their respective talents. They graduated and married. Over time a pattern developed. He worked more and more at his art; she worked less and less. They had a child. She moved even further away from her art.

Now she realizes she underestimated the strength of John's ambition and his need to succeed: "I gave him all the room and support he needed and he took it. But I was part of the competition. He did not think to encourage me." Most men are socialized like John. They are taught from birth that they must achieve, and that's how they live their lives. He did not worry about Patsy's career. Her taking on the home responsibilities fit in with his need to succeed as an artist. He did not question it. That was fine with him. But Patsy had never developed the sense of direction that John had all along. He knew what he wanted and went for it. She shared his ambition but did not want her work to interfere with the marriage. Yet Patsy never thought she would *not* be a successful artist. The more she neglected her work, the more she doubted her own ability and lost self-confidence. When her son was about two years old, the disappointment and unhappiness Patsy felt about her life peaked. She had fallen into a traditional marriage and had given up her work. But she never consciously decided to do so. It just slowly came about.

Luckily Patsy and John sought out professional help at this point and now recognize the choices they made by default. She felt early on John's need to succeed and his need to have her total support. Giving less to her art lowered the immediate conflict and, at the time, seemed a small enough price to pay for love. Patsy and John also acknowledge that it was easier for both of them to put her career "on hold" because she did not really know what she wanted to do. Patsy did not understand what she would be giving up: "I always knew I wanted to marry and have a family. But as talented as I knew I was, I never clearly saw myself as a successful artist." She does now. Whether the marriage can survive and their love for each other be rekindled are still up in the air. They both want it to, for their own sakes and for their son's.

Wives in dual-career families are generally more sensitive than husbands to the impact of their career pursuits on the family. One indication of this is that although they are much more likely than their husbands to press for the dual-career pattern, wives are also much more likely to relinquish or alter their side of it when stress or lack of support is experienced within the family relationship. Women often take on more of the parenting and thus experience much more conflict between the demands of parenting and work. Putting less into a career lowers the conflict.

Women also may fear the impact of personal success on their marriage and on their children's development. Women traditionally provided the emotional stability usually required for relationships to survive. More often than not, both partners still expect wives to provide this. If the husband in a dual-career marriage has an affair or the children develop emotional problems, the woman's career becomes the suspected culprit. Women typically get blamed for what goes wrong with families, and they learn to blame themselves as well. Stereotypically, they get the "bum rap" when problems occur with the family.

Consequently, some women relinquish career pursuits when husbands cannot provide adequate emotional support. Typically, rather than ask their husbands to provide more support for their professional work, women absorb more of the load at home and do what seems necessary to keep the peace at home. In the example already mentioned, Patsy decided it was not worth risking her marriage by continuing in a career that her husband found personally threatening. Other women cut back on careers or shift to less demanding positions.

A subtle double message gets directed at women today. On the one hand, women are expected to work. Families need two incomes. On the other hand, women are expected to put careers "on hold" or "drop out" of careers when children come along. The *Wall Street Journal* is among the first to tell us that "women are giving up the board room for the nursery." This psychological "out" provided by society puts women desiring a role-sharing marriage in a dubious position. They want to "share it all" with their partners, but they find themselves "doing it all," often by prematurely putting aside their initial career goals or lowering their sights because society gives women a latitude not extended to men.

Role Conflict and the Superwoman Syndrome

> Every time I come home from a trip, which is not that often, my daughter reminds me that her best friend's mother is always available to volunteer at her school and my husband tells me the refrigerator is empty. [A businesswoman in her late 30s]

Women in dual-career marriages want to develop professionally and establish a part of themselves separate from their husbands and children. And they want loving, close, connected relationships—marriage and children. Achieving this integration of separation from and connection to family without guilt and conflict, however, proves difficult. And for good reason.

Although our society has moved from the view that women lack the abilities to have careers, it has not moved from the position that women's nurturing responsibilities inhibit or at the very least affect their ability to be successful in their careers. Women are still viewed as the primary parent—by many women themselves, by the vast majority of men, and by society at large. A study of the Fortune 1300 Corporations, for example, indicated that 55% of the companies responding thought that parenting responsibilities are assumed primarily by the female partner.[14] None of the companies saw fathers assuming primary responsibility, however. Similarly, a recent article in the *Wall Street Journal* had the headline: "Executive Women Find it Difficult to Balance Demands of Job, Home."[15] We hear little about the role conflict and balancing acts of executive men. Although increasing numbers of professional men take an active part in parenting, few choose to stay at home with their children or cut back on their careers to devote more time to their families. In the majority of dual-career families, women do far more of the parenting and household work than men.

Not surprisingly, then, women in dual-career families experience much more conflict in combining work and family roles than men do. There are more demands on their time and energy because more is expected of them at home. But the demands of work remain the same for both partners.

One woman in middle management feels these pressures daily: "My boss expects me to perform like traditional men married to their job and my husband and daughters want me to be like traditional women without jobs." She and her boss, for example, frequently meet for working lunches; she grabs something from a vending machine and he feasts on quiche neatly packed for the microwave. She could, of course, pack a lunch for herself but rarely takes the time. With car pooling, kid's lunches, and last minute homework to supervise, she barely makes it to the office by 8:00. Her boss gets there by 7:45 at the latest, sips coffee, and peruses the stock pages. The contrast in their lifestyles angers her at times, particularly when he assumes her availability for Saturday morning meetings and she has a thousand other things to do, including going to her daughter's soccer games. Usually she goes to the office, feels guilty about missing the games, and squeezes in everything else.

Linda, the 43-year-old woman in this example, like many women in dual-career marriages, manages quite well despite the juggling and continuous conflict. She would prefer less stress and expresses resentment toward her husband when he fails to do his share of things at home. Overall, she sees some stress as inevitable when both partners combine work and family roles. She thoroughly enjoys her work and gives work and family pretty equal priority. She has a good relationship with her daughters, ages 10 and 12, and with her husband, and often talks over with them the conflicting loyalties she feels. They let her know in no uncertain terms when, from their perspective, the balance tips too much toward work, which it does at times. As for the guilt, she finds it both diagnostic and reassuring: "When I start feeling guilty about how I am juggling things, I take time out to look at my life. Am I not doing what I want to?" She also finds the guilt heartening: "Knowing I have a loving and supportive family makes it easier to immerse myself in work. But ambition is seductive. The more you do, the more you want to do. Guilt keeps me from going too far in the wrong direction."

Having a career traditionally means putting work first. Female socialization and societal views ask women to put family first. One common way for women to meet both these introjections is to become "superwomen." The prototypical superwoman strives to be a conventional wife, conventional mother, and conventional professional. She adds the role of career woman to her family roles and requests little of the family in accommodating her new obligations. Instead she expands her roles and makes all the accommodations, if any are to be made. Becoming a superwoman is the path of least resistance and provides a temporary solution, particularly when a woman gets little support for her career from her family. Limited time and energy, of course, ultimately take their toll and make this pattern unworkable as a long-term solution.

Lori, 36, well illustrates the superwoman syndrome. A highly energetic and talented woman with a doctoral degree in languages, she handles the roles of wife, mother, and lecturer with great aplomb. Her husband Ralph's position requires much out-of-town travel and Lori does everything at home—from the income taxes to diapers. Ralph could not be more pleased with the marriage and takes great pride in Lori's ability to juggle so many roles. Although at times she feels exhausted and ready to throw in the towel, Lori can't see herself doing things differently. All in all, she also feels pleased with the arrangement. Lori feels proud of her accomplishments, as well she should, and certainly does not view herself as a superwoman. No one is really at fault in this situation. Lori and Ralph set up a pattern that seems to work for them.

Living in a predominantly male culture contributes to women's role conflict and to the taking on of the superwoman syndrome as a way out. Women, not men, pushed for equal career opportunity. But must professional involvement come on men's terms? To succeed in careers, must women adopt the male attitude, placing work at the center of their lives and relegating relationships and children to a lesser role? Did women free themselves from the constraints of the traditional female role only to take on the constraints of the traditional male role? The world is full of men who basically estrange themselves from their spouses and children because of their careers; women with careers desire other options—a different reality—that the superwoman syndrome seems to promise. Family retains a high priority.

But superwomen, although they appear to be joining the ranks of women succeeding in careers, remain locked into the traditional position of adding their careers to their family responsibilities while their husbands' family responsibilities remain unchanged. Husbands continue to have little domestic and familial responsibilities beyond "bringing home the bacon." This situation ultimately puts married career women at a disadvantage professionally because they continue to carry the lion's share of family work. It also discourages many women from attempting or pushing for a two-career marriage because the societal message they hear is that women need to be "super" in order to achieve a career and a family. Combining a career with a family, commonplace for men, becomes for women a nearly impossible feat accomplished only by superwomen.

Not all women seek to reduce their role conflict by becoming superwomen, although many may try this strategy for short time periods. A time particularly conducive to the adoption of superwoman strategies occurs after the birth of a first child. Some women hesitate to share the baby's care and find themselves doing everything. This pattern does not bode well for the future, however. Patterns established early on are likely to continue, and husbands may never become active fathers. Husbands may hold back because they see parenting as the wife's role. They may feel left out, especially if the child nurses. At such times partners must find ways to share the experience, which may mean forcing themselves to go against their automatic reactions. One woman, sensitive to her greater comfort with children, encouraged her husband to care for their infant right from the start. In fact, she purposely left her husband "in charge" several times during the first weeks after birth so that she would not interfere with how he did things.

Other ways to resolve or reduce role conflict, which I discuss at length later in the book, require spouses to deal directly with the causes of the conflict. Typically, this means reconsidering attitudes about men's and women's roles and finding a way to share responsibilities at home that feels equitable to both partners. The key here is for both women and men to feel secure in what they do and to have little doubt that what they do is right for the couple and for the family. If, for example, a spouse constantly questions whether hours spent writing a book will harm their child's emotional development, the internal stress and role conflict will not abate. Resolving the conflict requires either deciding not to do the book or finding a way to do the book that leaves psychic energy and ample time for the child.

In fact, this is where many women and men may differ. Men do not doubt that they need to work and usually assume that working enhances their family, often regardless of possible costs to their actual relationships with their wives and children. Women struggle because for them family means interacting and feeling emotionally close, not keeping at a distance. I see this at times in my life with my husband. He has a deadline and blithely goes off to his office on a Saturday morning or spends the evening working at his desk. In the same situation I am more apt to agonize about whether I am spending enough time with my daughter, invite her to come to my office with me, or promise her something fun later.

* * *

Struggles to combine career and family largely stem from women's (and men's) experiences growing up and from resistances from the society at large. All too often women find themselves "doing it all" because that's the path of least resistance, at least in the short run. But "sharing it all," their eventual goal, requires changes in men as well and appears more and more likely.

CHAPTER 3

CAREERS AND CARING: CAN MEN MANAGE BOTH?

> There is no question that the rewards of sharing career, achievement, and childrearing with one's spouse are great, but the price paid can be high. It is a price that many of us raised as men never imagined we would have to pay, and therein lies much of the trouble. [One man's view of struggles with a role-sharing dual-career marriage][1]

Men's and women's lives are inextricably linked. What affects women affects men, and vice versa. The women's movement in the United States has been inspiring and empowering women for nearly 25 years. Women now experience greater economic freedom and personal opportunity than ever before. Traditional male-dominated professions such as law, medicine, and business report substantial proportions of women. The professional labor force, which was 26% female in 1960, claims 39% women in its ranks today. The majority of these women have husbands and most have children.

These changes in women's lives have had and continue to have enormous impact on men. Women, especially those in dual-career families, ask men to be partners in a very different kind of relationship. They want someone to look across at, not up to. Men are their equals, not their superiors or benefactors. Women who themselves expect to and do earn good salaries look beyond what a man can provide economically. Sensitivity, emotional depth, and companionship mean much more. Women's greater financial independence also enables them to leave a bad marriage. I was talking to a woman in her 50s at a party recently. She gave up her work as a nurse when she married over 30 years ago, has three adult children, and is casting about for something to do. She feels envious of women today. They can ask for so much more in a partner and expect to get it!

The cover of a widely circulated popular magazine recently declared that "the American male is new and improved." The article described an emerging male image combining strength and sensitivity, action and introspection: "a way to be strong and no longer silent."

A poignant example of this change in men emerges in a profile of a 36-year old, widowed history instructor in the book *Some American Men* written by former *New*

York Times reporter Gloria Emerson.[2] The instructor, who had been raising his eight-year-old daughter by himself since she was 15 months old, explained to a female student why he was not afraid to take a daring political step before being granted tenure. The 22-year-old senior expressed surprise at his response. She expected some statement about the importance of living up to principles or some reference to the sacrifices men must make for society. Instead he responded personally: "The worst thing in my life [the untimely death of his wife] has already happened. What is there to be afraid of?"[3] His emotional disclosure startled her and reminded her that some men were changing too. Relationships with women, in this case with a professional woman in a role-sharing marriage, take a high priority in men's lives, and many men willingly now let the world know this. His comments gave this student hope that she at her age might just live to see these kinds of changes in many men.

The student's surprise reflects the unusualness of the male instructor's responses. Some men are changing, but many still resist modifications in traditional male behavior and privilege. A recent comprehensive survey of "how men feel" about women's demands for equality and power provides some pertinent statistics.[4] On the one hand, over 85% of the men surveyed approved of their wives being employed. The percentage of men who "genuinely supported women's demands for independence and equality," however, paled in comparison—between 5 and 10%. The men surveyed recognized the economic importance of a wife's income but clearly saw it as a second income. They approved of women working out of necessity but not out of equality and self-sufficiency. Women's economic independence was valued and appreciated as long as it did not pose a threat to the traditional male position.

Traditional views about the rights and roles of men and women remain firmly ensconced in society. This continuing reality puts men in two-career families in somewhat of an incongruent position. On the one hand, unlike women, men have experienced few constraints in their traditional roles and have not asked for change. On the other hand, an egalitarian marriage promises benefits for male partners, although these may seem less dramatic and perhaps less evident than those for women. Men in two-career marriages report two gains in particular, both stemming from changes in the traditional male role: sharing the provider role and bonding with children. Historically, men took full economic responsibility for the family, and wives and children became dependents. Men in dual-career marriages, however, share the economic burden. Wives' incomes, for example, allow husbands to shift careers, go back to school, or work fewer hours. Men typically also removed themselves from child-rearing, leaving this to women. The joys of parenting were culled more from the success children achieved as adults than from actual relationships with them growing up. In contrast, many men in dual-career families desire close emotional ties with children and express their human needs to nurture and bond more freely because their marital relationships encourage it.

Although changes in traditional sex-typed views clearly benefit and enrich men's lives, they do not come without recalcitrance and resistance—both from the society at large and from within men themselves. Men today have few nontraditional role models and mentors; their fathers did not believe in sexual equality, nor did they present alternative images of maleness or teach them how to balance active participation in work and in family. Men, therefore, must explore new options and travel in unmapped terrain. This, of course, can be hard. Struggles inevitably occur even for the well-meaning male. As one man said[5]:

> Sometimes I think, "My father didn't do it that way, my friends don't do it that way, why should I?" But then I realize that those rules aren't in effect and it turns things from black and white to gray.

Growing up in "a man's world" profoundly affects how men think, act, and feel. My 13-year-old daughter once asked, "Why do the boys in my school think they are better than the girls?" Essentially we said, "That's how boys are raised but girls today know better." My husband explained how boys pattern themselves after the men they see in movies and on television and the fathers they live with. These men often assume they are better than women, and boys, like the men they become, take this for granted as well. They are brought up to believe that they have more and get more in life because of an inborn superiority. My husband added, "Boys worthy of your anger must possess the capacity to understand that their assumed superiority lacks substance. They were born into it and did little to deserve it." The explanation from her enlightened parents did not really satisfy her. Our daughter was likely too young to comprehend her dad's wisdom. But we both knew he said this for our benefit as well. Our own struggles over the years involved these very issues.

Entitlement and the Prerogative of Male Superiority

> I've heard complaints from many single women that they can't trust any man's willingness to promote their careers. Speaking for one man, myself, I think they are right. Men are too used to putting themselves first. [A 32-year-old male lawyer who feels ready to settle down]

Almost all societies view the two sexes differently, and in most societies males are more highly valued than females and are superordinate to them. The extent of this differential evaluation may vary, but its result is always to grant men greater power and privilege.[6] Examples abound. In China, where overpopulation poses grave problems, couples are permitted one child. Families desire a male child; if the first child is female, the infant's life is sacrificed. In India the burning of brides recently made the headlines. The husbands up the ante for dowries and "accidentally" burn their brides when the dowries are not forthcoming. Couples in the United States divorce less frequently if they have a male child. In addition, women here continue to face occupational prejudice and discrimination.

Thus, not unlike the adolescents in my daughter's school, men grow up with feelings of confidence and specialness simply because they are born male. This specialness is an essential aspect of male entitlement, which encourages men to feel that what they do or want should take precedence over the needs of women. One husband, a hard-working 3M middle manager, married to a hard-working accountant, still finds himself furious when his spouse gets home too late to prepare a hot meal for him and the family. It remains a bone of contention between them. He refuses to eat leftovers and hates casseroles. In defiance, he resorts to cold sandwiches, which he eats before she gets home.

Most if not all men in dual-career families struggle with feelings of male specialness. The struggle is a rite of passage. Husbands especially grapple with their sense of entitlement when wives push for an egalitarian marriage. Being the supportive husband or involved father a man wishes to be takes on a new meaning when so doing interferes with his own career goals or dampens his need to feel dominant. A highly ambitious and successful writer described his struggle to live with his ambition and stay in his dual-career marriage[7]:

> When I hold my daughter or brush out her hair or tell her stories, I am frightened by the side of me that wants to push ahead at her expense. At the same time, I feel that by taking on more of my wife's load I am losing ground in my career; it's like swimming with rocks in my pockets.

Male resistance comes from several sources. Many men advocate equality as an ideal but find when it comes to their own marriages they consider their wives' careers as secondary. Their professional success takes on greater importance than their wives'. Moreover, many men believe that a successful husband should not have to do housework—that's part of being a man, women do things for you. The failure of these men to provide emotional support to their spouses and their limited participation in family work reflect these views. As one man said, "It's totally unfair and unreasonable, but she does everything around the house because I don't cooperate. Rather than argue, she does it." Another husband commented, "Intellectually I think the old values are unnecessary and wrong, but in practice my wife takes on more responsibility for the house and children than I do." Both these men live in traditional dual-career marriages. Their wives pursue less demanding careers and take care of things at home. The men's resistance to helping out causes little marital conflict. Should their wives push for change, the situation would be quite different. Men with strong entitlement needs rarely can make the changes necessary to sustain a role-sharing dual-career marriage. In fact, many couples who eventually divorce do so because the woman feels patronized, and the man knows she is right.

Feelings of entitlement may also engender feelings of competition and resentment between spouses, which may have positive or negative effects on the marital relationship. On the one hand, an ambitious wife can enhance a man's own ambition and encourage him to move into new and creative directions. As one man said, "We nudge each other." At the same time, a man may be unable to get excited about his spouse's career success when he views his career advances and salary as lagging behind. I recall a conversation with a man who interviewed me for a radio broadcast on the experiences of men in dual-career families. After the interview he shared a personal incident, which had come to mind as he listened to me on the show. The previous evening, his wife, also in the communications field, had come home ecstatic. He was the first to know of her promotion and raise. He wanted to share her excitement, but found that he just couldn't—he worked as hard as she did, was as talented, and now she was moving ahead more quickly than he, although she was several years his junior. Until this incident, he had never thought that she might be the more successful half of the couple, and the anxiety associated with that possibility caught him by surprise. He had considered himself "a liberated man who was beyond that kind of nonsense." He admired his wife's abilities and had always been one of her chief supporters. But the night before, he had hardly slept, and he left the house that morning embarrassed by his jealousy and anger. We talked for some time. He called later to thank me. The label "male entitlement" helped him to understand what had seemed so irrational and crazy—why should he be so jealous of his wife when he wanted her to succeed? The self-understanding allowed him to discuss how he felt with his wife, although he still found it hard to share her joy.

Because both men and women are socialized to view the husband as the primary "bread winner," both spouses may view the husband who is less professionally successful than his wife as less manly or dependable and as not carrying his fair load. Embarrassment and sometimes conflict may arise for the man and perhaps his spouse as well. A man who finds himself in this situation may look for ways to maintain his "superior" position. For instance, a typical response for a man whose spouse earns more money is to make sure there is an understanding that the work he does is exceptional and requires a special talent, dedication, or loyalty to an ideal. He may earn less, but what he does has importance that cannot be measured in dollars. One man, a lawyer dedicated to social causes, makes a modest income. His wife's higher salary allows him this freedom. He feels uneasy about the situation and finds himself taking on harder and harder cases to prove the worthiness and significance of his work.

Some men could not care less if their spouses earn more. Others report conflict, particularly if the man feels pressure to do more at home—which in most cases is no more than a fair share. This situation becomes especially troublesome if the husband has a strong sense of entitlement. His wife's request may be quite reasonable—after all she puts in longer hours and does more at home—but he finds it insulting. Conflict between the partners also emerges if the wife needs the husband to be more successful than he is *or* than she is. However, this situation, which I discuss in a later chapter, rarely causes conflict when husbands do their fair share of the family work. Bad feelings and resentment come when husbands have less ambition and expect women to do everything at home.

Another implication of entitlement needs to be recognized: that men are to remain free of the domestic role. Women have pushed for egalitarian relations more than men and have expanded their roles to include employment and career opportunities. No one, certainly not the male establishment, would doubt the potential benefits. These include power, prestige, and money.

Men, on the other hand, are being asked to expand their roles to include family work. Here the benefits are less clear in the eyes of society and perhaps of the individual men involved. As one writer noted in an article about how men are coping with female success: "Women have begun to ascend—have ascended—into roles of power and prestige once reserved for men. No corresponding change *in acceptable roles* has occurred for men."[8]

This comment makes explicit what we all know. Although women's traditional work within the family has great importance, it has far less value than men's traditional work outside the family. And it has even less value when engaged in by men. Moreover, as the "politics of housework" makes clear, taking on roles traditionally assigned to women becomes problematic for men because they may begin to question their specialness and sense of masculinity. Few men willingly wash the dishes and many are threatened by associations with domestic activities. Stereotypically masculine men are not found in the nursery or kitchen. In fact, many men, including some in dual-career families, continue to view housework and child-rearing as "women's work" even when their spouses are employed full-time.

One last point with regard to entitlement needs mentioning. Men who support equality for women often do so out of an ideological commitment to equality as a concept or ideal. Their commitment is *not* part of their personal identity or consciousness. For such men, feminism is an abstract issue of rights and obligations, a set of rules to be followed. This perception can cause difficulties because the man's tendency may involve his deciding what marital or household obligations should be expected of him, fulfilling those obligations, and then expecting everything to be acceptable. He may think or even comment to his wife, "I washed the dishes and put out the garbage, how can you want anything else?" What the woman wants, however, is for him to put aside his own needs and wants and to view the situation from the perspective of what needs to be done to maintain the house and family. All too often men implicitly assume that that's women's work and they're being supportive simply by helping out. For example, a wife may well resent and be upset by her husband's expecting to be thanked for cooking dinner when she shops, vacuums, and washes clothes and no one really notices, since it is her job to do so. He feels a sense of entitlement and recognition, which she finds irritating to even have to respond to.

Power and Dependency

In many ways the relationship between women and men is a power relationship, with men having more power than women.[9] Yet, as Sempé captures in this cartoon portrayal of conventional marriage,[10] husbands establish a profound dependency on wives. Traditionally, men expected women to provide them unconditional warmth, sustenance, and love but saw this as prerogative, not dependence. Women needed men and took care of men in return—this was men's due. Just as female dependency on men became exaggerated, as I

"I'm cold, I'm hungry, and I want love!"

discussed in the last chapter, male dependency on women was understated. The traditional view that women are weak and men strong leaves no place for male dependency. Dependency and weakness become synonymous with the behaviors and emotional needs of women; independence and power become synonymous with the emotional needs and behaviors of men. Thus, to understand and recognize male dependency, a crucial issue for partners in dual-career families, one must look first to power and then to men's socialized needs for power over women.

Power

Power typically refers to the ability of one person to cause another person to behave differently than she or he would otherwise. Men have more access to resources that are likely to help advance their careers, a crucial element to their acquisition of greater power. Even now, despite women's vastly increased opportunities, more men than women receive advanced degrees and hence obtain greater knowledge. It follows that a greater variety of positions are available to men than to women, and the positions filled by men typically are accompanied by higher salaries and greater prestige.

Many men base their self-esteem and life goals on traditional values that associate manliness with dominance, power, and aggression. Successful men in our society act independently and do what is necessary to make their way in the world. To be "real" men, and successful men, they must be effective breadwinners, whatever the physical and emotional costs. Because manliness, or even personal identity, gets tied to the "good provider role," men often become controlled by their occupational obligations and society's standards and symbols of economic success. For example, a man whose spouse returns to work at midlife because his income cannot keep up with increasing expenses often reports lowered self-esteem. If the wife is very successful and brings in a good income, the husband may feel anger and resentment at his loss of authority in the family and loss of the breadwinning role. One man in this situation said he no longer had a role to play. He had lost his special position with his family and felt pretty worthless to them.

Some men who consciously choose a dual-career situation report similar anguish, but for different reasons. They put all their energy into being successful and expect to be applauded for their single-minded pursuit. Instead, wives prefer more of a partnership and want husbands who share themselves with the family.

Many men look to their breadwinning role and how much money they make, not to how successful they are as husbands and fathers, to confirm their manliness. Manliness becomes associated with being more successful than wives and having wives attend to husbands' needs to succeed. Furthermore, a strong man requires a weak partner, namely, the woman who depends on him.

Dependency

No one questions men's needs for power; men's needs for dependency, in contrast, rarely get mentioned. Not surprisingly, then, men's dependency on women becomes evident as a need for power over women rather than as a need for bonding between equals. In essence, male dependency stays hidden behind a cloak of power and dominance over women.

Several writers propose reasons for male dependency remaining hidden.[11,12] They see society as the culprit. One popular theory describes boys as bribed by promises of power and dominance to prematurely relinquish any dependent tendencies.[13] Parents consider boys from a very early age to be strong and aggressive and raise them to be fearless. Boys don't cry. Boys are mommy's little men. Boys know what to do. Boys never admit fear, and so forth. Because boys must behave like "little men" before they can comfortably do so, according to this theory, they may depend more than they realize on unconditional acceptance and nurturance from women to sustain their pseudoindependent stance. Thus, men depend more than they realize on women's responsiveness to them. Women's need for and caring for men keep them strong. (Similarly, according to this theoretical view, society bribes women with promises of love and approval to prematurely abandon independent tendencies.)

Joseph Pleck, a psychologist who studies men's changing roles, identified three psychological sources of men's needs for power over women.[14] Each, as we shall see, relates to male dependency but expresses itself as a power that men attribute to women, which in turn leads men to dominate women.

The first of these, called "masculine-validating power," concerns penis power. Through sexual relations with women, men look for validation of themselves as men. The close association between sexual power and manliness pervades American culture. Men and women alike hear the same basic message—in sex, as elsewhere, it's male performance that counts. The man must be big where it counts, powerful, take charge, and successfully orchestrate sexual activities. Moreover, sex serves as a vehicle by which men gain status with other men. In describing his youth, one man in his 40s said[15]: "What I learned about being a man . . . was that women were game, an erection was a weapon, penetration clinched the bargain of power and ecstasy. Pleasure was incidental; one seldom spoke of how something felt, but only of potency conveyed."

Male partners reporting a loss of sexual interest or problems with erections often feel relatively powerless in the marital relationship. For some men this involves unexpected failures at work or disappointments with their careers. For others the dysfunction or disinterest mirrors struggles for dominance in the marriage, which the husband associates with threats to his potency. For instance, the wife may be demanding that her husband participate more in housework and parenting or seriously be considering an attractive move that will require relocation of the family. To effectively deal with such problems, partners must recognize the unconscious connection men typically make between dominance and sexual functioning. Then solutions to the particular event precipitating the crisis can be found.

In one case of a role-sharing couple, the wife became increasingly disgruntled with her job opportunities in their present location and wanted them both to look for positions elsewhere. Her husband, who owned a small business, felt very conflicted. He agreed that her situation looked grim, yet he felt extremely satisfied with the locale and his position. At some point during these discussions he began having erection problems and the couple sought counseling.

Interestingly, the couple's egalitarian history obfuscated his obvious knee-jerk dominance response. He basically opposed moving and wanted her to put his needs first. The more she talked about relocating, the more powerless and defeated he felt. His reactions to her wish to relocate, however, were not apparent to either of them and did not emerge until they were both in counseling. When he recognized the connection between male prerogative and his being upset, he became less resistant to the move. His perceptions shifted from needing to dominate to examining what would be best for both spouses. When the couple began considering the pros and cons for both partners, the inability to get erections ceased to be a problem. The couple, by the way, did eventually relocate.

The need for power over women, then, reflects men's dependency on women to make them feel like, or be, men. One 32-year-old husband in a dual-career family said that he felt so much more of a man after he found out that his spouse was pregnant. He went on to say that he did not think he needed women to validate his masculinity so much as he sought vehemently for concrete evidence of this masculinity within himself and within his impact on the world: "Women seem more acutely perceptive of that which is masculine than I seem to sense. I know that this is like giving women power, but at the same time I wonder if women are socialized to be more perceptive on these matters than I was." This husband, not unlike others, projects onto women power they don't really have. Some men also project onto women their own dependency on women. They see wives as wanting too much from them when in fact they themselves desire a great deal of emotional support and understanding from women.

Men also cede to women "expressive power," the power to express emotions. Women often express men's feelings for them and draw them out. Men can be "soft" with women—they become safe havens for male emotion. This aspect of male dependency continues to be a crucial part of marriage. Husbands in dual-career families readily admit that wives often give them much more emotional support than they can give in return.

One husband said his wife does it so naturally that "it's second nature for her and I just expect it." Numerous women in my studies echo the comment of one lawyer: "I complain about a bad day at the office and my husband says, 'Quit and take a different position.' That response makes me furious. It cuts off any conversation and shows little empathy and interest in me. I never say that to him. I expect him to have bad days and he expects to tell me about them in detail."

Such situations relate to a third way men depend on women. Women play a part in men's power with other men: Women reduce the stress of competition between men because men can always be superior to women. For example, one sure way to put down a man is to tell him that even a woman can do better or that he behaves like a woman. This dynamic emerges in dual-career families if the husband expects to outperform the wife but the reverse occurs; then he feels demoralized and angry.

The struggle to achieve a healthy balance between passive dependency and autonomous strivings represents a lifelong task for women and men alike. In fact, helping men and women develop and integrate both independent and dependent strivings formed an important component of the 1960s women's movement. The less widespread men's movement of the 1970s also emphasized how important it is that men recognize their own dependency and value it positively as an integral part of male development. Men were encouraged to validate their own self-worth and that of other men instead of needing women to do this. Increasing numbers of men today value emotional expression in themselves and within their intimate relationships and feel prepared to be nurtured and to nurture others. But here again some worry that the sword is double-edged. Will men who develop their "softer," feminine side be admired from afar, but be deliberately avoided or held suspect because they deviate too far from the male norm?

Nurturance

> The "sensitive man" is what women today think they want, but in reality women consider such a man a "wimp." And men know it. [A man recently divorced after eight years in a dual-career marriage]

The vast majority of men learn from birth that nothing is as masculine as money, power, and achievement. And nothing should get in the way of this—certainly not women or attachments to children. In order to acquire power, prestige, money, and other perquisites of achievement, men are asked to sacrifice emotional expressiveness and intimacy in their relationships. Society discourages men from being fearful, timid, sensitive, and needy. Men, in turn, become reluctant to express emotions or dependency, and when they do, it is typically with women, not men.

The ability to be a part of a mutually confirming relationship is difficult for many men. Often they remain unaware of the depth of their dependency on women for nurturance, support, and maintenance of their own feelings of well-being. In addition, their sense of entitlement can get in the way. They may feel they should be treated as special and that their concerns are inherently more important than their partners'.

Mutual spouse support entails valuing a spouse's abilities and ambitions as well as a willingness and ability to provide emotional support, empathic listening, and nurturance. Putting aside one's own needs to be nurtured and emotionally sustained and doing so for the partner takes on special significance, as does drawing out the emotions and feelings of the other person, something men typically depend on women to do for them.

A well-known anecdote illustrates the stereotype of men's woeful inability to nurture[16]:

> A golfer returns home late from a game and explains to his wife that his companion, Fred, caused the delay. Apparently, Fred had a case of sunstroke and collapsed on the second hole. The man shakes his head sadly and says to his wife, "It was absolutely awful. For the next seven holes, it was hit the ball and drag Fred, hit the ball and drag Fred."

In this illustration we see in exaggerated form a man's inability to put himself aside and evaluate the situation from another's perspective. The golfer's concern with his own needs causes him to ignore Fred's dilemma and to see Fred as making his life harder.

This not very flattering view certainly describes very few men in *successful* dual-career families. Mutual spouse support remains the single most important ingredient for an enduring dual-career marriage.[17] Although many men report difficulties initially, they learn over time how to provide support to their partners. This learning requires good communication and good will between the spouses. Wives may at first feel resentful about instructing husbands on how to be supportive of them. This passes quickly, however, if husbands make earnest attempts to listen to what wives want and need.

Men attracted to the idea of a dual-career family often envision close involvement with children. Indeed, one of the motivations of men who choose the dual-career lifestyle is the desire to nurture—particularly their own children.[18] The reward for them involves the traumas and joys of the emotional ties we all have to bear for love. Close emotional relationships have increasingly become a core part of these individuals' sense of self as men. A shift of values seems to be occurring in our culture toward men's greater participation in parenting. One bit of tangible evidence of changing attitudes is the decision by officials at an airport in Connecticut to "change with the times" and install diaper-changing tables in men's restrooms.

Fathers provide on the average about a third of the child care when mothers are employed full-time. Approximately 11% of preschool-aged children whose mothers are employed full-time receive care primarily from their fathers.[19] In a recent study of 78 dual-career families that I conducted, 14% of the husbands did more parenting than wives, 32% pretty much shared the parenting, and 54% did less parenting than wives.[20] (All these families used some outside help such as paid day care.)

Yet some disinterest and discomfort still remain. Some men fear that becoming too close to their children will result in career hindrance or loss. In consequence, they forestall possible conflict by remaining distant from their children. Other men just don't consider fathering all that important. Evenings and weekends devoted to work provide much more satisfaction. These men respect and encourage their partners' brilliant juggling of career, marriage, and parenthood, but they feel no need to try it themselves.

* * *

Many men who desire egalitarian relations with women discover, to their surprise, that such relations generally require them to do more than simply share their status and power with women. They must also grapple with how to nurture spouses and children and share with spouses the conflicts and joys of combining family life and career pursuits. Often they must revise their sense of self and find new sources for their self-esteem and masculinity. This is not a one-time task, nor is it easily accomplished. Difficult as it is, some men successfully engage in this process of change and find it personally rewarding as well as essential to a happy and thriving marriage.

PART II

LIVING AS A TWO-CAREER FAMILY

The day-to-day life in a two-career family converges around the personal relationships between the partners, the life of the family, which usually includes children, and the couple's two careers. Each adds a unique component to the marriage and brings its own set of rewards and struggles.

CHAPTER 4

THE HUSBAND–WIFE RELATIONSHIP

> If partners are to remain emotionally alive in enduring relationships, they must dare to shake up established patterns—to strive for new goals and to dream new dreams that involve each other.[1]

Love and Marriage

A recent theory sees love as a triangle.[2] The three sides of this love triangle consist of passion, intimacy, and commitment. Passion, the motivational side of the triangle, leads to physical arousal and an intense desire to be united with the loved one. "Love at first sight" and mutual feelings of attraction, which reflect the passion dimension of love, often provide the impetus for a romantic relationship. Passion alone is rarely sufficient, however, and a second element, intimacy, must develop for the relationship to last.

Intimacy is the emotional aspect of the love triangle and includes such things as closeness, sharing, communication, and mutual support. This dimension of the love triangle manifests itself in a number of ways. Emotional intimacy, for example, involves comfortably expressing innermost thoughts and feelings. Intellectual intimacy centers around discussions and dialogues or even travel pertaining to professional work, current events, or mutual interests such as history and architecture. Recreational intimacy reflects a shared interest in leisure activities such as gardening, skiing, hiking, jogging, or traveling.

The cognitive side of the love triangle is expressed in commitment, both a short-term decision to love another person and a long-term commitment to maintain that love. Levels of intimacy, passion, and commitment change over time, and so do relationships. Consummate or complete love—the love on which we base marriage—ideally contains high levels of all three elements, whereas romantic infatuation includes only passion and long-term friendship only intimacy and commitment.

This love triangle may oversimplify a complex and subtle phenomenon. Nonetheless, it highlights important aspects of enduring marital relationships:

- Giving and receiving affection
- Giving and receiving help
- Understanding the other person and feeling understood
- Sharing personal ideas and feelings
- Sharing ideas and information
- Receiving and providing emotional support
- Allowing and enhancing each other's personal growth
- Making the other feel needed and needing the other

Certain aspects of an enduring relationship take on a special meaning for the dual-career family. In the traditional marriage, the wife supports the husband and children emotionally and does all she can to sustain and enhance the husband's career. Her sense of identity and purpose comes from the marriage and the roles she performs within the marriage. The husband draws his sense of identity from his career and the financial support it provides for the family. Thus, the meaning of intimacy and commitment is in many ways determined by the roles taken on by wife and husband. She needs him to support her financially and he needs her to take care of him physically and emotionally. She gives him the room to do what he needs to do to advance personally and professionally. He entrusts to her the needs within the family.

In contrast, many women today who are involved in a career—and perhaps men to a lesser degree—enter marriage with much less clearly defined roles. Women, not men, push for the dual-career pattern. Women want the opportunity to develop and define themselves apart from marriage and therefore seek relationships with men who share and support this view. Men, on the other hand, are not that unhappy with how things used to be. To them, it's nice to have a second income, but why can't everything else stay the same?

Often partners do not know what they really want or what would be satisfactory. However, to weather a dual-career marriage, partners must be willing "to shake up established patterns," to live with ambiguity, and to learn as they go. Thus, showing a partner that you need him or her and feeling needed in return takes on a new meaning in marriages in which both partners have good incomes. Needing becomes less an exchange based on expected roles and more an interdependency between two people based on what needs to be done, time constraints, and mutual understanding.

Allowing one another to grow takes on special significance for the dual-career couple. Often this need for growth becomes manifest when career opportunities somehow infringe on the other partner. For example, one of the biggest struggles for women in dual-career families is taking the space they need and not feeling guilty about it. As one woman said, "After our daughter was born it was terribly important to hear my husband say I am good at what I do and the family does fine when I work late." Valuing and taking joy in a partner's strength, talent, or advancement, a sine qua non in dual-career families, falters, however, if a husband (or wife) feels threatened, less important, or less valued because of it.

Are Two-Career Marriages Egalitarian?

> It's your day to lick the puppies and my day to bury the bones.[3]

Myth has it that everything is equal in dual-career families. He cooks three days and she cooks four days; the next week they switch. She gets the car serviced even months and he does it odd months. A nanny raises the two children. There are no conflicts and there is lots of money.

Nothing could be further from the truth. Perhaps most true of dual-career marriages is the lack of congruence between the egalitarian ideals of the partners, particularly those of wives, and the reality of how they live their lives. A relatively small percentage of couples are what one would call equal.

Equality in marriages is typically determined on the dimensions of power and role specialization. Power refers to the extent to which one partner dominates the other in the marriage. Similarly, role specialization reflects the extent to which sex determines who does what to maintain the marriage. Traditional marriage links benevolent male dominance with clearly specialized roles assigned on the basis of sex. The husband makes the major decisions, and his career is preeminent. The wife cooks, cleans, and does the laundry, and the husband pays the bills and mows the lawn. Egalitarian marriage rejects this traditional view of power and role specialization, although in general not completely. In reality, most dual-career marriages achieve a middle position.

The degree of role-sharing achieved by couples most likely reflects attempts to maximize joint successes in the face of social realities. Men still command higher salaries than women and have greater upward mobility. Women are still viewed as the primary parent, as evidenced by accommodations provided by employers, such as maternity leave and the location of child-care centers; few men willingly interrupt their careers for families. And we cannot ignore the continuing influence of sex-role socialization and male prerogative.

Still today the woman assumes a greater proportion of the responsibility for family work than the man and more likely subordinates her career interests to those of her partner.[4] Even among couples in which the wife is a high-level professional, household tasks are often divided in a sex-stereotypic manner and wives perform a greater percentage of the household tasks.

Husbands' involvement in family work often takes the form of "helping out" rather than taking over responsibility for a particular activity. Often when a husband contributes to domestic work, it is with the understanding that he is operating in "female territory," and he more likely undertakes such tasks jointly with another family member than alone. Overall, however, men's family work is increasing while women's is decreasing. A recent survey of American couples[5] reports that among couples who held egalitarian views, approximately half the women spent less than 10 hours a week on housework, one third from 11 to 20 hours, and one tenth more than 20 hours; approximately three fourths of the men devoted less than 10 hours and one fourth from 11 to 20 hours.

This represents a move toward convergence, although it is far from parity. What role overload remains may be due more to women's feeling that there is still inequality between the partners than to their feeling overwhelmed by the demands of multiple roles.[6] In other words, it is not so much that women feel they are doing too much as that they perceive that men are doing too little.

Characteristics of Spouses in Two-Career Marriages

Personalities and Values

> I love your husband. He's very male without being macho.[7]

This comment most typifies men in enduring two-career marriages. Their sense of manliness does not stem from a need to dominate women. Rather, these men possess high ego strength and emotional security and a capacity to value their wives' independent identity. Some men enter the marriage with this self-view; for others it evolves as the relationship becomes more egalitarian.

As I mentioned in the preceding chapters, male and female socialization still inculcates traditional notions about the sexes. Even men and women who hold fairly egalitarian views on entering into marriage find themselves confronting stereotypic notions of male entitlement and specialness and female nurturance. Modifications in attitudes and behavior may be necessary for the relationship to endure. Partners most willing and able to make needed changes possess high self-esteem, a commitment to both their careers and the marriage, and an openness to altering things for the greater good of the relationship. Typically, they acknowledge liberal views about the rights and roles of women and men, although wives tend to hold more liberal views than men.

Characteristics evolve for husbands and wives alike in mutually supportive marriages. Women often show a stronger sense of self-direction, independence, and confidence. For men, interpersonal warmth and expressivity take on more prominence. In essence, partners encourage one another to display characteristics inhibited by traditional socialization. This mutual enhancement of expressivity in husbands and of self-esteem and independence in wives may be an important ingredient in the marital quality of successful dual-career marriages. Neither partner may have a need to dominate the other and both may have the skills to communicate.

Men with high needs for power and traditional views of sex roles make unlikely candidates for a dual-career marriage. Highly competitive men with high needs for achievement show little interest in supporting wives' careers. Because competitiveness also inhibits the spontaneous, vulnerable, and self-disclosing behaviors integral to close relationships with a child, participation in child-rearing also would be unlikely. Several studies describe the husbands of career women as reflective and intellectually oriented, rather than power- and competition-oriented. Liberated husbands are responsive, sensitive men who take their wives' career needs into account and who make some effort to be reasonable if a compromise must be made. These men's self-images are less determined by traditional notions of masculinity that require a wife whose life revolves around her husband.

At the same time, high levels of ambition in either spouse may make a two-career marriage impossible to sustain. How much ambition a marriage can afford cannot be determined in advance, but common sense suggests guidelines and limits. Individuals whose career goals hold far greater importance than anything else clearly would be unwilling to make compromises or allow room for a second career.

Inclusion, the desire to contain or enclose another within the ambit of one's self, is another important personality attribute. In some ways inclusion refers to the romantic aspect of marriage—two people becoming one person. Traditionally, however, only one partner, the wife, depended upon the other for self-definition. In contrast, both women and men in dual-career marriages often cite the benefit and importance of maintaining separate and independent identities. Both partners value a separate sense of self-directedness and an ability to fulfill oneself through one's own interests, quite apart from their intimate, sexual relationship. Thus, wives and husbands in dual-career families describe themselves as relatively more self-reliant and self-sufficient than partners in more traditional marriages. One husband, a businessman in his 40s, finds his wife's relative independence liberating: "We both have things to do and undertake them freely, knowing that the other's happiness is not totally derived from the relationship."

A common problem for spouses in dual-career families is finding time for each other. Prime time typically gets allocated to children and work; time for the relationship inadvertently gets a much lower priority. But time together, and high-quality time, are essential to marital happiness. The greater the importance of an intimate relationship to an individual, the more effort must be put into maintaining the marital relationship. Despite their relative self-reliance, we find that the relationship is "paramount" and "all-important" to many enduring dual-career couples. The comment of a couple married 15 years reflects a typical sentiment: "It's the difference between feeling loved and accepted and secure on a daily basis and not feeling that way." These couples perceive marriage as enhancing, not limiting, their lives.

Affiliation also predicts relationship satisfaction. Affiliative needs refer to the desire to connect or associate with others. Spouses with moderate to high affiliative needs approach marriage with a spirit of accommodation and cooperation. They attempt to understand and support each other and ensure time together. Some couples reserve a certain day for lunch or dinner together. This time is sacred and always takes first priority. Other couples plan early breakfasts before the children get up or occasional weekend trips.

Individuals with relatively high needs for affiliation also value associations with their own friends—outside their marriage. This can pose a problem if partners feel time with friends means time taken away from them or from the family. Friends, however, bring a vital force to relationships. They often pursue a similar lifestyle and bring different perspectives to identical concerns. Moreover, we look to friends for impartial feedback about our thoughts and behavior. When I feel out of whack, I sometimes call one of two good female friends, both in dual-career marriages. They usually listen intently, and may offer advice. More often than not, though, I just need a sounding board, and because of our long-standing friendship, they know they are to provide it.

Mutual Supportiveness

> There are few sociological laws better supported in the research literature than the law of husband cooperation.[8]

A husband's support for a wife's career goals is central to marital success. Women traditionally supported their husband's personal and professional goals; it was understood that men needed support. Men, in contrast, are not socialized to assume supporting roles. Many men, even those advocating equal opportunity, have difficulty giving their wives the emotional support and encouragement they need in their careers. A woman in her early 50s, for example, was furious. She would come home with complaints about the office, and her spouse, a highly successful banker, would say, "You don't need to work. Quit." True, they could manage quite well without her income, but that was not the point. This husband thinks he *is* being supportive; he does not like seeing his wife upset. But she hears, "Your work is not important." She wants him to take her concerns about the office seriously, to respond to her as she would respond to him in this situation, and to stop patronizing her with suggestions that in effect say, "Big Daddy will take care of you."

The level of spouse support is also a matter of male prerogative. Although many men see their spouses' careers as important, they also see them as secondary, and the support they offer reflects this view. We live in a society in which relatively few *married* women hold positions of power and leadership, although many may have interesting careers. Men are well aware of their spouses' abilities and ambition; they just don't take their careers as seriously as their own.

Individual men differ on how much change they can tolerate with regard to traditional sex roles. Some men delight in their wives' career success and might be the first to tell colleagues that their wives earn more than they do. Others get upset and, as I discuss in a later chapter, feel threatened and resentful. Irrational feelings of competitiveness often emerge. The expectation of wives in this situation—that their husbands will share their good fortune or carry their fair share at home—only aggravates matters.

Attitude–Behavior Consistency

> I feel like I married a traditional man in nontraditional clothes. When we were living together everything was shared, we never even had to discuss it. Now he expects me to do everything. I resent it terribly and we can't even talk about it. I love him very much but I am afraid our marriage is not going to survive. I can't be a superwoman. [Young professional woman who has been married less than a year]

The greatest challenge facing dual-career couples involves fundamental changes in attitudes or priorities. Despite claims to the contrary, many dual-career couples unconsciously or even consciously accept conventional assumptions about men and women and the nature of marriage. Often these assumptions remain unspoken until after marriage or until spouses are placed in a position in which they are expected to act in accordance with supposed egalitarian beliefs. A common complaint among women in my studies is that many men talk a good egalitarian game, but when it comes time to put their words into action it's another story. They expect women to take the back seat and to support what they want.

Men are not the only culprits, however. Women, too, play a part in alleged male resistance. Women tend to marry men who are older, wiser, and more successful. Few career women select less ambitious men as partners, and when they do, friends and relatives question the soundness of the decision. Although the times see it as more acceptable, only a minority of women marry younger men—although women outlive men by about ten years on the average. A tacit alliance becomes established in which the man, more than the woman, *needs* to be successful for psychological and societal reasons; his career should take precedence.

Rather than confront or challenge husbands, wives often hold themselves back careerwise and/or absorb more of the load, perhaps unconsciously considering this inequality as fair or nonthreatening to the husband in the long run. One woman received an excellent job offer at another university and asked her husband to consider relocating. He said that would be fine. He had no objections, but he wanted assurance that she felt she could be as successful in her new position as he would be in his present position, if they stayed in their present locale. She turned down the offer but was furious with him for months. She later realized that she wanted him to make the decision for her; she feared asking too much of him and risking his happiness. Rather than discuss the matter further, she declined the opportunity out of hand.

Resistance to change may come as a surprise to well-meaning spouses, yet they are entirely predictable. Change is hard, especially when it concerns a tradition as long-standing as roles assigned on the basis of biological sex. In some cases even nontraditional partners feel hesitant to act on their convictions. A critical period is after the birth of a couple's first child. Time and again, the couple feels adamant about shared parenting. They both attend Lamaze classes and he assists with the birth. But somehow she ends up with the lion's share of the responsibility. He does not hear the baby cry and she is reluctant to wake him for 2 A.M. feedings. She does not like how he bathes the baby and he feels criticized, et cetera. Spouses must talk at this crucial point because patterns may form based on resentment and resistance rather than on mutual agreement and choice. Understanding why they are unhappy with one another and how their relationship strayed so far from their ideal is essential to the success of the relationship.

Combining Work and Family

> Highly educated women realize that it is not necessarily their job to take care of household tasks and the nitty-gritty details of parenting. I try to do as much as I can and it stays pretty even. [A scientist in his late 30s married to a woman with a career in law]

How to combine occupational and family roles presents a thorny and continuing source of stress for both women and men in dual-career families. Conflicts between work and family are intrinsic to the dual-career situation and present choices that parents must face. However, neither gender has role models for doing so, and neither gets much encouragement or assistance from his or her professional world. The traditional societal norm has not changed—the family is still expected to accommodate to work demands. Work comes first.

Equity vs. Equality

A sense of fairness or equity about the balance of family and professional roles is critical to the success of dual-career marriages. Equity refers to the fairness perceived by individuals in a particular situation. It rests on a comparison an individual makes between his or her own cost/benefit ratio and that of another in the same situation. Indeed, individuals may evaluate a relationship or situation as equitable even though it entails an inequality of conditions. Thus, the concept of equity involves subjective judgments of fairness.

In contrast, the concept of equality involves relatively objective measures. It is based on the notion of dividing everything down the middle, whereas equity allows for preferences—perhaps for one person doing all of one task and none of another.

What matters is the outcome, not the process. Do both individuals feel that fairness prevails?

Clearly, equity carries the danger of applauding women for how far they have come while not allowing them truly equal opportunity. In essence it could reinforce traditional role relations within a marriage. At the same time, equity also provides for a degree of flexibility. We might question how much progress would be made if a rigid system of traditional sex-role assignments were replaced by an equally rigid system of egalitarian role division. Moreover, the 50–50 sharing of egalitarianism is likely a chimera. For practical and psychological reasons it is simply not attainable. Very few dual-career families are egalitarian in the true sense of the word.

At this point, the successful working out of a dual-career relationship rests largely on women's ability to press for and achieve what they consider an equitable situation, although to an outsider it might appear far from equal. Generally, women take on more and give up more than men in attempting to make this arrangement work. This appears temporary, though, because increasing numbers of men and women hope and fight for equal costs and benefits and envision "sharing it all" as their eventual goal.

Patterns of Role-Sharing

Generally speaking, men and women approach family responsibilities differently. Men's traditional role makes legitimate their placing career over family, while society downplays possible costs to relationships with children. For the male spouse, retirement or children's departure from the home often brings an awareness of what he as a father may have missed. Women's traditional role, in contrast, makes legitimate their placing family over career, while society downplays possible costs to their personal development and fulfillment. Many women, for example, worry that emphasizing their careers will endanger relationships with their husband and children. At the same time, they also worry that giving too high a priority to their family will jeopardize progress in their careers. For them, the decision not to have children or the departure of teen-age children may bring relief from role conflict and permission to devote themselves as much as they would like to their careers.

The extent to which *both* spouses agree to a particular marital pattern, regardless of what pattern they choose, predicts marital satisfaction. If both spouses agree to a traditional male-dominated pattern or a role-sharing pattern and live this agreement out in their day-to-day interactions, marital quality will be high. Problems come up when the spouses do not agree. Most often such problems occur because the husband is more traditional than the wife.

Basically, there are three dual-career family patterns: traditional, role-sharing, and ambivalent (see the first table). Couples who agree to one or another of these three patterns generally do not differ in marital happiness. Where they differ is in what they bring to and want from the relationship. These differential factors, summarized in the second table, are illustrated in the following comparative descriptions of each family pattern.

In the first pattern, which I call the *traditional dual-career family,* the responsibility for family work is retained by the woman, who adds the work role to her traditionally held family role. In these families, the man believes that work within the home is women's work. The woman in these families generally accepts and acts upon the same premise.

Typical Ways in Which Two-Career Couples Handle Work and Family Roles

Two-career family pattern	Household responsibilities	Parental responsibilities	Career responsibilities
Traditional	Mostly wife	Mostly wife	Both spouses
Role-sharing	Both spouses	Both spouses	Both spouses
Ambivalent	Mostly wife	Both spouses	Both spouses

Factors That Influence How Couples Combine Work and Family Roles

Personal factors

- *Personality:* How important is a person's need to dominate, have an intimate relationship, be emotionally close with children, be number one in her or his field?
- *Attitudes and values:* What are a person's beliefs about who should rear children, who should be breadwinners?
- *Interests and abilities:* How committed are the partners to their work? How satisfying are their careers? How successful are they?
- *Stages in careers and life cycles:* Is one spouse peaking careerwise and the other opting for early retirement? Are both spouses in the initial stages of careers? Are both spouses at midlife but one established careerwise and the other just beginning?

Relationship factors

- *Concepts of equity:* What seems fair? How do spouses come to this agreement?
- *Spousal support:* To what degree can one spouse, particularly the husband, put his or her needs aside and do what is needed to support the other?
- *Shared values and expectations:* To what degree do spouses share life goals, views of men and women?

The spouses also agree that the husband's career is the preeminent one. Knowing that this is the case, the husband feels free to support the wife's dual roles and to genuinely appreciate her efforts. But the wife is more supportive of his career efforts than he is of hers. A husband in his early 40s, for example, described his wife as "the best of the old and the best of the new. She does everything around the house because she enjoys it." Far more professionally ambitious than their wives, the men in these families typically earn significantly more than their wives and see their wives as highly successful in combining their careers with family life.

Spouses in this pattern often see the choice of a dual-career lifestyle as "her decision." They understand that the wife could work only in the home if she wanted to and the family could adjust to living on the husband's higher income. Typically, partners in traditional dual-career families were both still in school when they married. Entering the marriage before entering careers likely influences how the spouses view roles and how they divide family work after careers and family are established. The pressures of two emerging careers, for example, may force spouses to give one career a greater priority, most likely the husband's. Since women typically get less direction and encouragement than men about professional goals, when push comes to shove they also may absorb more of the domestic responsibilities. Many women interrupt their education or career pursuits at this point and become full-time homemakers. At some later point, when wives reenter the work force, patterns for handling family and home responsibilities, which grew out of a traditional arrangement, become difficult to change.

At the other extreme, and approaching the ideal image of the dual-career family, is the second type—the *role-sharing dual-career family*. In these families, both spouses actively participate in both household duties and parenting. The role-sharing dual-career family is clearly the most egalitarian of the three types and best represents the pattern that many couples strive for.

Spouses' salaries tend to be comparable, as does their involvement in child-rearing and household chores. The men in these families are not simply "helping out" but rather are what some people call "the modern male." There seems to be an implicit assumption that neither spouse is more responsible for home responsibilities: "When one of us gets busy, the other takes on more responsibility."

Of the three family types, spouses following this pattern are most apt to discuss *before* marriage the career expectations of both partners, how household and family roles are to be handled, the importance of children, and so on. Many also grapple with possible ways of finding two positions in the same geographic area and handling time commitments so as to assure time for the marital relationship. Also, mutual spouse support for career and home responsibilities is highest among these couples, although husbands at times perceive that they are more supportive of their spouses' career efforts than vice versa.

Some women in role-sharing marriages admit that they struggled in the early years of the relationship. They feel they had to keep themselves from doing more than their fair share and to remind husbands that they had career demands too. Because men expect women to accommodate them, they sometimes lack spontaneous awareness of what needs to be done. They come home and automatically sit down and relax with the paper. They go to the grocery store with a list and that's all they purchase—even though they know milk is needed but somehow missed the list. Over time, husbands change and become aware of what needs to be done. At this juncture wives stop policing things and trust that husbands will follow through.

The third type, called the *ambivalent dual-career family,* may be a transitional one. Here, as in the role-sharing dual-career family, parenting is shared by the spouses. However, the woman retains primary responsibility for maintaining the home and doing such tasks as cleaning, cooking, and shopping. Husbands in these families are actively involved in the day-to-day rearing of their children and very much want to be. A close, intimate relationship with their children is paramount. They do not want to father the way they were fathered and end up estranged from their own children. They entered the marriage with the conviction not to do so.

But for them, housework is another story. Intellectually they think the old values are unnecessary and wrong, but in practice they expect wives to take care of the home and the details of day-to-day household responsibilities.

Typically, these men show little motivation to change and seem content to live with a somewhat ambivalent situation. Spouses' salaries and commitments to careers appear quite similar, although wives tend to be more supportive of husbands' careers than the other way around.

The genesis of all these ways of combining work and family is complex. Generally speaking, the husband is typically more involved in family work when the wife contributes more financially *and* when greater meaning and importance get attributed to her work.

The smaller the difference between the husband's and the wife's income, career aspirations, and spouse support, the more involved the husband is in domestic responsibilities, particularly parenting. Overall, for spouses in a role-sharing or ambivalent dual-career marriage, stress is minimized by mutual spouse support, willingness by the man to actively engage in parenting, and comparable career opportunities in their present locale.

* * *

Critics and even supporters of the dual-career lifestyle may err in assuming that spouses in dual-career families are self-centered "me" people who want to have their cake and eat it too. Often overlooked or minimized are the stress and compromises inherent in working out two successful careers, a successful marriage, and care for the children all in the same family. Role expectations, personality characteristics, and mutual support are all related to the kind of role definition or redefinition achieved by couples who live in what remains a male-dominated society.

Dual-career marriages "make it" despite the external constraints because the two individuals know they want careers and a marital relationship based on all three sides of the love triangle—passion, intimacy, and commitment. One couple married for 15 years and very much enamored of each other captured the spirit of these marriages in their comment: "We see the whole as stronger and more vibrant when the two parts are independent and strong."

CHAPTER 5

WHAT ABOUT CHILDREN?

> The world is full of fathers who rarely see their children [because of their careers]; for mothers [with careers] to do the same thing is not the answer.[1]

Children pose by far the most troublesome issues for partners in dual-career families. For some the greatest anguish concerns whether to have a child at all. For others it's where they'll find the time and how they'll manage all the day-to-day parenting activities.

Dual-career couples' anxiety about parenting makes sense. Women today have options other than having children as well as greater control over what direction their lives take and what brings fulfillment. Traditionally, careers and children were mutually exclusive. Careers belonged to men. Men in turn counted on women to devote themselves to the needs of children. Most women, even those educated for careers, accepted that view. It's different today. Both women and men in dual-career families expect to continue their careers and to derive great satisfaction from their work after they have children.

Most women with careers who marry expect men to share in parenting. But men's participation in child-rearing remains a rather revolutionary practice, perhaps even more resisted by our society than women's commitment to their own careers. A decade of research on fathering clearly indicates that men have the capacity to care for their children competently, yet parenting is still equated with mothering in our society. Few men see themselves as having to choose between a career and fathering, but many view their wives as facing that choice.

One husband in a dual-career marriage commented that perhaps deep down he was never fully convinced by the feminist movement[2]:

> Having a daughter has made me re-examine my own attitudes toward women's issues. As I hear about the tribulations of single women, childless couples, and infertility, I often wonder who my daughter will marry, or if she will marry at all. Will she have children? Will she follow in her mother's footsteps, successfully, brilliantly juggling a career, a marriage, and motherhood? Or will she opt for the role her mother now

> often considers, staying at home to be with, nurture, and teach her children? I tell myself I don't care whether Annie has a career or is a housewife. But it does matter to others.

This husband, who genuinely admires and supports his wife's many roles, discovered that he still primarily associates dedication to children with women. His wife, not he, experiences conflict between roles and may need to choose between them. The birth of his daughter has had little influence on his choices.

Traditional views about "who is the primary parent" are not entirely unrelated to feelings of male entitlement. Men expect to be successful at careers and to outperform women. Women in dual-career families often do much more of the parenting than men, and the women's careers are affected more by parental responsibilities. The upward mobility of husbands is rarely hampered by run-of-the-mill home responsibilities. This continuing reality for women and men alike affects decisions made with regard to children—whether and when to have them and how to care for them.

Moreover, because both partners pursue careers, even when parenting is shared, couples need assistance with parenting from people outside the immediate family. Unlike more traditional couples, dual-career couples who consider having children must also face the additional question of who will help them. Nannies and extended families belong to a different era and culture. Even spouses' own parents often live hundreds of miles away. This need for continual help with children counters tradition and causes resistance to role-sharing marriages. Spouses must grapple not only with changes in their own attitudes about how to "properly" rear children but also with finding suitable help in meeting the needs of their growing children.

Making the Decision

> We haven't decided not to have children but we wonder whether our desire to have children warrants the sacrifices that would be involved. [A couple in their late 30s who love their work and share several strong outside interests]

Some dual-career couples struggle with whether to ever have a child, and most if not all struggle with when to have a first or another child. A child dramatically alters one's life, changes priorities, and requires enormous expenditures of time, energy, and money. Although most partners grow up thinking they will have a child, when dual-career couples consider the reality of what's involved, ambivalence and indecision can set in. Couples want to make the right choice for the right reasons; they don't want to be parents who have children they don't enjoy or didn't really want.

Women agonize over the parenting decision much more than men. Women typically do more of the parenting and worry more about how to integrate children's needs with career demands. Images of women as mothers run deep in our culture and choosing not to be one runs counter to views of what it means to be a woman. Some women fear making the wrong decision for the wrong reasons. If they decide not to, will they regret it later? Women face a time constraint that at some point makes the decision irreversible.

But the decision to parent, although it may appear to be the woman's, really belongs to the couple. The decision must be a *joint* one. One woman struggled and struggled with the child issue when she turned 36. Should she? Shouldn't she? She often described her dilemma to her husband, who was more reluctant to have a child, and presented him with the pros and cons. Finally one night he blew up at her: "Don't I have any say in this? Just because I already have a child [from a previous marriage] and have less desire, don't I figure into the decision?" She immediately saw how right he was. In fact, should the female partner express a strong interest or desire to have a child, the male partner would be remiss if he were merely to go along with the decision, thinking little would be expected of him. More likely than not, she expects him to share in the parenting.

One couple, married seven years, had always had the understanding that they might not have a child. Then, when the wife was 39 and the husband was 47, she began to have doubts about the earlier decision and talked seriously about having a child. He wanted to support whatever decision she came to, but terribly resented being put in this situation. He preferred to remain childless, but did not want to have undue influence on the decision. At the same time, he knew if they did have a child, it would be with the understanding that they would be partners in parenting as they were in the rest of their marriage. He explained his predicament and she saw the truth in what he said. When she had imagined a child in their life, she had simply assumed he would totally share the experience.

On further discussion and reflection, she recognized that her desire not to miss out on something wonderful grew out of the strengths of their shared relationship. Their mutual respect, love, and support made her want to extend their experience together to include having a child. This couple debated for over a year and eventually decided to remain childless. A child, although important, would have meant devoting their middle adulthood years to caretaking. In particular the husband would have had to spend his 60s raising a teen-ager. Although the idea of having a child was exciting, the reality was less sanguine. Once they could look beyond the joys of loving a child, which was strong for both of them, they were able to acknowledge that was not how they wanted to spend their remaining years together.

As this example illustrates, what partners decide greatly depends on what meaning a child has for them personally and for their life as a couple. The age of the partners, their investment in careers, and their way of handling home responsibilities provide the context in which the decision eventually gets made.

If both partners always assumed they would be parents, and discussed this before marriage, the decision essentially becomes one of when to parent, which we'll consider later. Stress and conflict emerge when partners *differ* either in the importance they place on having a child or in views about rearing the child. How much conflict arises and its effect on the eventual decision often depend on the age of each partner, which spouse wants a child more, and the type of role accommodations characteristic of the marital relationship.

Age can ease or exacerbate matters. For example, partners close in age who marry in their 20s or early 30s and feel firmly committed all along to child-rearing may consider postponing a child until their late thirties as a viable option. Increased life expectations for women and men make the prospect of parenting at midlife quite palatable. On the other hand, individuals with no previous children who marry when the woman is in her mid- to late 30s may feel pressure to hurry up and have a child, even though the relationship may not have yet developed the stability to weather the stresses of childrearing.

Potential problems can also arise when one spouse, typically the husband, is 5 to 10 years older than the other. In this situation, illustrated in the example given earlier, the husband will be well into the last phase of his life when the child is reaching maturity. If both spouses feel strongly about having a child, the age difference has little bearing. But many men who marry or remarry in their 40s have devoted their 20s and 30s to career demands and now feel ready to ease up some and focus on other interests. Taking on additional family may feel more like a burden than an exciting opportunity.

Probably most conflictual is the situation in which the husband wants a child more but chafes at sharing home responsibilities. Some men feel strongly about having a child but are not that keen on committing themselves to shared parenting. Women who are very committed to careers and who want an equitable sharing of parenting duties feel reluctant to put themselves into a situation in which they will end up doing all the parenting.

What typically tips the balance one way or the other for these couples is whether they can reach certain agreements about parenting and can trust that both will faithfully fulfill their commitments without undue resentment or resistance. Take the somewhat traditional example, Mary and Jim. Mary always wanted children, and when she turned 30 she felt it was time. Her husband, Jim, wanted to wait a few more years. He worked long hours and just didn't want to devote the substantial energy to parenting. Their discussions revealed that she felt ready to cut down on her work and devote considerable time to parenting. This was fine with Jim, and they went ahead with the plan.

Mary and Jim's situation reflects one common pattern: The wife expresses stronger interest in and readiness for parenting than the husband and they both view her as absorbing most of the family responsibility. The decision again causes little conflict because Jim feels confident the rules won't change later. Whatever contribution he makes to the parenting responsibilities will be appreciated.

The converse, of course, is less true. If, of the two partners, the husband has the greater desire to have a child, he rarely expects to or volunteers to do more of the parenting. Even if he states a willingness to do most of the work, the wife may feel she still would run too great a risk. Admitting a low desire for child-rearing, and then actually remaining relatively nonparticipant, is easier and more socially acceptable for a father than for a mother. Women, for example, more readily receive such benefits as maternity leave and a flexible work schedule. Schools call mothers first. Thus, the wife may feel pressure to do more than she wants to because society sees her as the primary parent. In some cases the wife may also question whether her husband unconsciously shares that same view and, in effect, is making promises he won't keep.

Chris and Bill are typical of this situation. He very much wanted to have a child but she was reluctant. Bill had a stronger sense that family means children too. Chris gave more of herself to her work than Bill did. They shared roles at home, and he had every intention of doing more than his share with children. Still Chris feared getting overwhelmed by parenting. She knew from friends' experiences that men often underestimate the requirements of parenting. They discussed at length what he imagined her doing as the mother and what he saw himself contributing. He looked into paternity benefits and considered possible changes in his work schedule. Then they seriously pondered how child care would be handled and came to an agreement they both felt they could live with. Bill would do more of the parenting, and Chris felt confident that he really would.

Deciding No

The media tout a new acronym—*Dink*—Dual-income, no kids. Dinks are a variation of Yuppie.

Deciding to remain childless is unusual. The media would have us believe that many couples today decide "no" and represent a growing proportion of dual-career couples. In reality, although many couples consider "no" an option, relatively few *decide* to remain childless. Specifically, about 3–6% of couples choose to remain childless, and this overall rate has stayed about the same over the last 30 years.[3] Most dual-career couples have one or at most two children.

It's hard to know who decides not to have a child. Such decisions remain private and rarely become the subject of general conversation. I, of course, being curious about these matters, often wonder why a particular couple is childless. Inquiring becomes tricky because while some couples actually make the decision, others evolve into it, and still others have it made for them. Partners may postpone a child only to find that when they finally decide to have a child, getting pregnant or carrying the fetus successfully proves impossible.

For the majority of couples, the decision to be childless is not made before marriage or even early in the marriage. Rather, the decision comes after the couple develops a lifestyle and vision for their future life that they are reluctant to give up. By far the most mentioned reasons for choosing to remain childless include freedom from child-care responsibilities, greater opportunity for self-fulfillment and career development, and a more satisfactory marital relationship. Couples who gave these reasons chose alternatives to life fulfillment other than child-rearing and reported satisfaction with their lives. As one couple said: "Our marriage is stable and enduring most likely because neither of us persuaded nor dissuaded the other to accept a childless lifestyle. Rather we came to the same decision, albeit probably not for the same reasons."

Typically, the female partner first considers not having children and mentions this as a possibility. The husband picks up on this and supports her choice because he too has little ego-involvement in having a child and little interest in devoting a significant portion of his time to child-rearing. A lawyer in a role-sharing marriage was not against having a child, but she knew she was too engrossed in her career to devote a lot of herself to parenting. Her husband realized he would have to assume primary care for the child, and although he would have liked to have a child, he was not willing to take on more than his share. He felt sad in some ways, yet when he considered how full their life was and how happy they each were, he knew that not having a child was the best decision.

Another couple who decided to remain childless realized that parenting would fall primarily on the wife's shoulders. The husband was not interested in actively helping to raise children and was often out of town for his work, which he thoroughly enjoyed. The wife did not see the rewards of parenting justifying the increased demands it would place on her. A highly energetic and devoted person, she derives great pleasure from her work and also tends to most of the details at home because of her husband's travel schedule. Not unlike many other childless couples, the wife in this family also earns more than the husband, and their joint incomes allow for "a very comfortable life."

For some couples, then, the preference to remain childless evolves out of a marital relationship of some duration and the reality of their professional lives. It represents the endpoint of a process of postponing a child because the time never seemed right for the wife to take on parenting and the husband was not that eager. Both spouses in these couples tend to work in high-status positions and to have very high levels of commitment to their work. Their passion is for each other and they greatly respect and support each other's careers. They feel devoted to one another and sense the importance of the relationship. One couple married 20 years said, "We looked at our lives and how we spent our time and decided whatever time we do have, we want to spend with each other."

A second common example of childless dual-career couples involves remarriage. Typically, the husband has children from a previous marriage and the wife does not. Husbands who remarry tend to be more professionally established than wives and considerably older as well. Should the wife enter the marriage with mixed feelings about having a child, the husband typically does not encourage having one. Because the husband has already fathered children, he prefers not to further complicate their lives with additional children, particularly when he is at midlife or beyond.

Probably most distressful for partners is when one spouse feels greatly opposed to having a child for reasons having to do with career and lifestyle. The partner who feels differently may fear that pushing the other way would put the marital relationship in jeopardy. In the case of Sam and Chris, further aggravating this situation was Sam's mentioning before marriage a preference for a childless lifestyle. Chris honestly thought she preferred not having a child but over the years began to change her mind. Sam remains adamant. They can pick up and leave whenever they want; they both travel a lot and a child would totally disrupt this arrangement. She basically agrees but fears giving in to his wishes and regretting it later. In this particular case, time will likely decide for them.

Deciding Yes—and When

> There is never a right time. You just reach the point where the reasons for putting it off are not all that compelling. It's not that you are ready. You are willing. [A lawyer and mother of two who says her children make her better at what she does]

Deciding to have a child brings the next challenge—when? Most couples take longer than expected to "finally get around to having a child" because they keep waiting for an opportune time in their careers. That certainly was my and my husband's experience. Our daughter arrived nearly ten years after we married. Interestingly, most people thought we had decided against children even though I was only in my early 30s at the time. As little as 10–15 years ago this "late age" was considered somewhat dangerous.

In contrast, most dual-career couples today delay parenthood until the female partner reaches her 30s. This gives partners time to get established in their professions and to take on the demands of parenting at a time of greater financial stability and emotional maturity. Moreover, contrary to earlier thinking, medical experts assure us that healthy women in their 30s or 40s enjoy a good prospect of giving birth to a healthy infant and remaining well themselves. Because fetuses with major defects can often be identified with amniocentesis and other methods of prenatal diagnosis, it is not until past the age of 45 that risks to mother and baby begin to climb significantly. The risk of an abnormal birth for women between the ages of 35 and 44 can be reduced to about the same very low level that it is for a woman of 34.

Because women bear the children and do so with less risk before reaching their 40s, couples understandably feel pressured to start a family once the female partner reaches her 30s. The wife's age usually determines the limits to postponing. In fact, the vast majority of women over 30 years old who give birth are in professional careers. Women in their 30s now account for one birth in four.[5]

Women wait until their early 30s because that works out better.[6] As I mentioned, waiting gives them a chance to mature personally and professionally. Thus, they feel less trapped by children because they have established a sense of personal and professional identity and have their careers in place. Waiting also puts them in the position to expect husbands to share more in child-rearing responsibilities.

The biggest risk for wives who have children at an earlier age is that they may have to forfeit career advancement. Many women who give birth in their early to mid-20s often interrupt educational programs or never really get themselves located in careers; as a result, afterward they find themselves unemployed or working in positions not particularly rewarding, financially or otherwise. These women typically take on more and more of the parenting and household work because they lack direction in their careers. Inadvertently, patterns characteristic of a traditional marriage get set up and later become difficult to change.

Delaying pregnancy, particularly past the age of 35, can pose risks, of course. Nature does not always cooperate, and the first child may come along less quickly than was hoped or perhaps not at all. Also, prenatal diagnosis, while reassuring, can mean selective abortion. This is emotionally difficult for some couples and not acceptable to others. Delaying, however, seems less hazardous than jumping in too soon. Couples need to evaluate what a child means to the wife's and husband's careers and plan "when" from there. For most couples the best time is when both partners feel established in their careers and the marriage rests on solid ground both emotionally and financially. If these things are not yet in place, waiting a year or two would likely be better.

"When" also arises with regard to possible second or even additional children. Not unlike many other couples, my husband and I found this decision more difficult than deciding to have a child in the first place. We always knew we wanted a child but never seriously considered how many. Part of our problem came from the idea that "children" means at least two. You can't have just one. An "only" child conjures up images of an overprotected, neurotic child who grows up alone in the world. Luckily, I suppose, we happened on an article in *Psychology Today* on just this topic and realized the extent to which we had swallowed a stereotype.[7] Only children fare as well as other children. They grow up to be normal, healthy adults; in the areas of achievement and accomplishments they perform much like firstborns in families with more than one child, probably because they get the special attention new parents shower on the first child. As adults, they often take on leadership roles and expect more of themselves.

Although few dual-career couples initially intend to have an only child, despite increasing acceptance of this in our society, a sizable portion never get around to having a second child. The reasons vary. In my own case, my husband and I both hoped for a girl and got our wish. (Obviously, couples who prefer and get a son may stay with one child for the same reason.) We both grew up in families consisting predominantly of sons and desired a change from that. Also, we both saw the 1970s as a superb time to have a daughter. The successes of the feminist movement led us to hope for her the same future that earlier generations reserved only for sons. Had our first child been a son, it's possible we would have decided to have another child. Then again, we probably would have been equally enthralled and would have felt—as we do now—that the child we have is everything we ever wanted.

For others, the urge may be strong, but having a second child may get outweighed by other factors such as the financial cost of two children and the time investment. Age can become a factor. Delaying the birth of a first child until the wife is, say, 35 or 36 often leads couples to see additional children as too risky. Also, couples may not want to extend their active child-rearing years into their 50s.

Parenting in Two-Career Families

> It warms my heart and melts it when my son snuggles up to me. [A 30-year-old father in a role-sharing marriage who manages a family business]

Fears about what will happen to the children in dual-career families are greatly exaggerated. A common scenario depicts two self-absorbed parents obsessed with promoting their own interests, while their children are only conceived and borne by the parents and thereafter, essentially parentless, are raised to adulthood by a series of strangers.

Reality finds children in dual-career families very much loved and well cared for. Most couples feel ebullient about being parents. For some it is better than they had imagined. Once the decision is made, relatively few couples see the rewards as not worth the commitment of time, energy, money, and emotion that go into raising children. Partners devote themselves to their children and consider parenting and a sense of family highly important. Wives see mothering as something they should want, and do want, to provide their children. Husbands vary more in their perceptions of fathering—not of its importance, but of actual participation. Indeed, men who choose the dual-career lifestyle often desire to nurture and relate closely with their children.

Dual-career families commonly have a broader view of parenting than the one typically adhered to. They agree that children need to feel cared for, secure, and loved, but believe this can be achieved in various ways. The bonding between mother and child, while of critical importance, is not the be all and end all. Bonding between father and child gets as much emphasis, and both are considered crucial to children's development. The basic notion these couples hold is that "you're supposed to have two parents in your life when you're growing up."

A second difference must be noted. Traditionally, women related closely to others while men did the self-developing. Both women and men acted on the supposition that direct caretaking by fathers was either innately impossible or unnecessary. Women bonded with children and put aside their separate sense of self; men developed personally and professionally and put aside connecting or relating to others. Women seemingly merged with children; men remained noninteractive, aloof, and distant. In contrast to such views, which assume that one parent, usually the mother, must give up a certain sense of self to administer to the needs of the child, the emergent dual-career view is that children benefit when parents respond to the child's needs without having to give up a crucial part of their own personal identity. As one father in his late 30s in a role-sharing marriage noted: "Our son might be better off if one of us stayed home with him. But we are healthier people because we work. This will benefit our son in the long run."

This, then, is the crux of the "new" parent–child

relationship that partners in dual-career families envision and strive for. Parenting means mothering and fathering and requires supplemental care. Moreover, each partner develops a viable balance or integration between giving and nurturing, on the one hand, and maintaining and developing one's own self, on the other. One father who delights in his role-sharing marriage emphasizes the importance of this integration for himself:

> For me it comes down to identity. If I were to stay home full-time with my daughter I would feel lost somehow. Part of the reason I am comfortable in my role as a parent is that I am secure in my role as a professional. Without that, I would not feel good enough about myself to be the kind of father I am.

Patterns of Role-Sharing

Despite partners' commitment to share child-rearing, a sizable proportion struggle with achieving this goal and some fall into traditional patterns. Their difficulties relate to tradition and stereotypes. Few spouses in dual-career families assume that women are innately better equipped to raise children, but many still believe that children need their mothers more than their fathers and that men should do less than women in the home arena. The society in which they live certainly holds these views. Also, male and female partners alike must learn how to integrate giving to and nurturing children with their own needs for self-development. The potential difficulty couples may encounter in integration cannot be overemphasized. Couples need to know that resistances may emerge, and often do emerge, because changes required go against deeply ingrained stereotypes of how women and men behave. Ways to combat such resistance are discussed in later chapters.

The various parenting patterns that result range from the role-sharing partners strive for to the more traditional patterns in which wives do much more than husbands. When asked who and what shaped their participation in child-rearing, most husbands mention their wives and the influence of the women's movement. Women are generally the key factors in change, both as instigators and beneficiaries of shared parenting. One man, now in his 40s, is the youngest of three sons reared in a traditional, college-educated family. From an early age he knew he would marry a career woman: "It always struck me as so ridiculous that girls were invariably at the top scholastically and then expected to marry, have a family, and forget the rest. I saw myself as different. I wanted to marry a career-minded woman and share things at home. But I am not sure I would have really made the effort if my wife had not insisted. I am very thankful now, although there were times I resented the intrusion on my life."

Husbands who participate more fully with their wives place greater importance on being a parent than do husbands in dual-career families that adopt a more traditional pattern. Parenting for both spouses in role-sharing marriages stays on a par with their career pursuits. Daily schedules get arranged to allow ample time for their children. A child's early years cause parents the most difficulty. Not only is suitable help sometimes hard to find but also parents may want to care for their infants themselves. Role-sharing partners often look for ways to integrate parenting into their work lives. I had a crib in my office when my daughter was 4–7 months old. She was not with me at work all the time, but only at times when I knew I would not be in class or meetings. Over the years I noticed that a number of my colleagues, male and female, made similar arrangements. This kind of plan, of course, requires considerable control over one's schedule. The only individuals I know of who have been able to do so comfortably own small businesses or teach at universities. The new publisher of Random House made news when she asked to have a nursery installed across from her office because she was expecting her first child.

Some role-sharing couples work out arrangements for managing most of the child care between them until their child is six months or a year old. These parents work at home some portion of the week and employ a home-care person to assist them with the caretaking duties the times they are away. One role-sharing father arranges his work schedule so that he can care for his six-month-old son during the afternoons. He is a full participant in all that needs to be done—from diapers and formulas to doctor's appointments. His wife stays with their son until 10 A.M. and then takes him to a home in the neighborhood. The caretaker, a woman licensed by the state to provide care for up to three children, tends to their child the remaining hours of the day and during other times when their work schedules conflict.

This couple finds the arrangement quite satisfactory although somewhat hectic at times. The husband, a professional writer, works evenings after his son goes to bed. The wife, a dentist, works until 6 P.M., which pleases her patients. They see this arrangement changing as their son gets older, but it works for now. Most importantly, they each get "private time" with their son. "Private time" for this couple means the opportunity to give their son their total attention some part of each day and to relate to him with as few distractions as possible from domestic or work responsibilities. Nurturing a child takes time and parents must give this time, not only in the early years, but all through childhood. Although no formula tells parents how much time is the "right" amount, parents know in their hearts when they neglect children and fail to give of themselves.

"Parent co-ops" provide still another option for role-sharing parents. In fact, dual-career couples often instigate their formation. Many parents worry about the quality of day care. Being a member of a co-op can give greater control over the care provided. Parents typically participate in the program's development, donate time to work in the center, help with fund-raising, or even serve on its board of directors. Perhaps better known than most because of media coverage is the Senate Employees' Childcare Center, located a few blocks from the United States Senate Office buildings in Washington, D.C. A number of senators place their young children there.

One couple, now parents of a 9- and a 12-year-old, knew from the time they took childbirth classes that child-rearing would be a joint venture. The only area in which she consistently takes more responsibility is clothes shopping for their children. Another parent, a role-sharing father of a 13-year-old, regularly jogs with his daughter, drives her to and from school, attends school-related activities, helps her with homework, and has season tickets with her for various sporting events. He describes their relationship as very close—"as close as I could have imagined. I know there are some things she talks to her mother about first, but that's normal for a teen-age girl."

In contrast, partners in traditional dual-career families consider the wife as the main and more important parent. The wife often remains the psychological parent who worries about what needs to be done and how to make sure it gets done. She usually earns less than the husband, takes on more responsibility for children, and, as a result, often subordinates individual career interests. The husband spends less time on a regular basis with their children and makes little effort to rearrange his schedule to accommodate them. Generally, however, the men seem comfortable with the amount of time devoted to parenting. A case in point is a highly successful scientist in his early 40s who is very content with the relationship he has with his two-year-old daughter: "She knows that Daddy has to work long hours and goes out of town a lot. We make the best of the time we do have together." In this family pattern, then, female partners show reluctance to relinquish aspects of the traditional female turf and husbands show reluctance to take it on.

Supplemental Care

Some form of child care is the norm for over half of American children, and this situation is not likely to change. But locating quality day care remains a persistent and recurring stressful task. Concerns about the general unavailability of quality child care is pervasive among dual-career couples. Although some corporations and communities provide some assistance, by and large individual families must obtain suitable day care on their own.

One can track the tone of the times through daily comic strips such as "Doonesbury" by Garry Trudeau, in which Joanie, the resident feminist, has it all except decent child care. Following a series of bad experiences with day care, she came upon a child care center that was "the answer to a parent's prayers." The ratio of children to staff was 2 to 1. All the care-givers, including the music and ballet teachers, had Ph.D.s, and none had less than 15 years' experience. There was a pediatrician on the premises, a psychologist on call, and a staff nutritionist to prepare lunches. The rate was $400 a day.[8]

Although this experience is fictitious, it illustrates how affordable day care can seem nonexistent or impossible to find. And for some it truly is. Although most parents report high satisfaction with the arrangements they work out, they view themselves as "extremely lucky." In some major American cities, such as New York and San Francisco, preschool care can be hard to find and competitive to enter. Waiting lists are not atypical, particularly for infant care.

The type of child care generally used depends on the age of the child. Most parents prefer group care for children older than three years, but they show no clear preference for individual or small-group care for children under three years of age.[9] Many families also ask parents or relatives to help for varying periods of time. Grandmothers often lend a hand during the first months after a child's birth even when they live far away. Relatives in the same city get called on to transport children during the day or care for them on the occasions of unexpected illnesses during the work week.

Some parents worry about the effects of day care and may need reassurance about their decision to use day care. Day care can have highly beneficial effects, especially on children's social development. Generally, children appear to profit from a verbally stimulating environment in which the adult caretakers and children frequently interact. Programs of higher quality typically have an experienced director and a relatively low child-to-caretaker ratio. For infants this ratio is no more than 3 to 1, and usually less than 3; for older children the range is 5–9 per staff member. Also predictive of higher overall care quality are experienced, trained staff, low staff turnover, and ongoing staff training.

Parents need to remember that the adjustment of children reared in the "traditional way" varies greatly, and the same is true of children who spend time in day care. As I mentioned earlier, there is no perfect way to ensure perfect children. Parents do the best they can in selecting supplemental care and in providing loving care at home. In fact, the focus of recent research on the effects of child care reflects this changing view. Studies now tend to center on the quality of care provided in day care rather than on comparisons of preschool-aged children in day care with peers in traditional homes.

The overriding question is how differences in the care provided affects children in day care. Just as parents' own care of children varies, so does day care.

Also crucial is the combination of parental care and other forms of care. If the day care is excellent but parents just are not available to their child, the child will likely be affected. The optimal combination, of course, is quality day care and quality parental care. Typically, parents seek out caretakers who share their own values of child-rearing as well as their general social values. For example, some parents seek out church-affiliated centers, while others eschew them. Parents interested in cognitive and intellectual development often place their young children in Montessori-type environments that teach children to read and reason from a very young age.

Strategies for Finding Suitable Child Care

The most effective initial step in considering child care is thinking ahead—and in several areas. Couples should first carefully assess their values and implicit assumptions about how they want to rear their child and what this implies for other aspects of their lives. For example, some spouses won't consider child care until after the first year because they want to spend this time with their child. Should this be the case, decisions need to be made about how full-time parenting can best be shared and what modifications in work schedules would be required. Negotiations with employers may be called for, or a rethinking of work schedules. Couples who plan to use supplemental care soon after a child's birth would need to focus less on negotiations with employers and more on locating caretakers whose philosophy and values with regard to parenting accord with theirs.

Some recommended steps are these:

- Gather information. Spouses should talk to friends and colleagues and inquire about what they did. The experiences of others can be invaluable. Ask them what they looked for in child care and why they settled on the care they did. Get the names of good care-givers. Inquire about areas in which you have reservations or fears. The responses of others may set these somewhat to rest.
- Check with employers about current policies and inquire about flexibility in the system. What exactly are their maternity and paternity policies? Could arrangements be made to work at home part-time? Many workable plans have never been proposed or tried. Creativity and perseverance pay off.
- Call or write city or state agencies such as Departments of Human Resources and obtain lists of licensed day-care providers. Read available books on what to look for in selecting child care.[10]
- Plan ahead. My husband and I lined up infant care when I was two months pregnant. Our daughter was registered at birth for a university nursery school. Planning ahead gave us great peace of mind. I know others feel the same way.

Afterschool Care

Day care pertains to preschool-aged children, the age group that immediately comes to mind in discussions of supplemental parenting. What happens after children enter first grade? The term "latchkey children" refers to school-aged children who come home to an empty house. Because most elementary and middle schools now provide afterschool programs, this problem appears less widespread. All kinds of summer programs are available as well, many of which provide full-time instruction or activities.

Perhaps of more concern these days for parents in dual-career families is what teen-agers do after school. As one 45-year-old mother of a 14-year-old noted: "I thought it would get easier as he got older. Instead, since he is in no one's care much of the time after school, I feel greater pressure to be at home and available—just in case. Yet I know he is old enough to be by himself or with friends for a few hours."

When children attend high school, and even junior high school, parental responsibility for their care is less direct and obvious but nonetheless still exists in part. Many teen-agers have little free time because of extracurricular activities, jobs, and homework assignments and would laugh at the notion of planned day care. Others, however, may lack interest in school, get involved in drugs, or feel little sense of direction—all with potentially destructive consequences.

Parents cope with this situation in various ways. Problems tend to be less frequent when parents show a strong interest in what their children do after school, hold them accountable for being where they say they will be, and show sensitivity to the difficulties of being a teen-ager.

What Happens to the Children?

> With great pride I announced to my four-year-old daughter that mommy had arranged her schedule to be home with her every afternoon. At first she cried. Then she said it would be O.K. if Mary, Jimmy, Suzie, Kim,... and Mrs. Wrightman could also come. [A college professor in her mid-30s very conflicted about not being home with her child]

How well do children in dual-career families fare? Are they harmed by day care? Are they selfish and self-centered? Is it too soon to know? The available evidence shows that children in dual-career families are more independent and resourceful and have a wider range of role models than children raised in more traditional homes. They benefit from greater contact and interaction with both parents and less exposure to sex-role stereotypic behavior in the home. Those from role-sharing families report greater self-direction and a closer relationship with their fathers than peers from relatively traditional households.

Children in role-sharing families also do more around the house and accept some fair share of the domestic responsibilities. A colleague of mine aptly noted that many parents don't take the time to teach children what needs to be done and so end up doing it themselves. She and her husband painstakingly taught each of their two children, when they were around four years old and still thought it was fun, to vacuum, set the table, cook, make salad, and so on. Now, five or more years later, she reports that "it's a routine. We all pitch in so we can have more time together."

Children see both parents doing what they want to be doing and thus being happier and more content when they are with the family. Says a university administrator in her early 40s married to a history teacher and who has two sons, ages 8 and 10: "We both work, we both want to work, and that's how it's going to be. Our philosophy is that children have to adjust to you just like you have to adjust to them. I am not worried about how they will 'turn out.' We have our ups and downs, but things generally work out."

What do the children say? In a recent survey of male and female adolescent and young adult children (aged 13–24) raised in dual-career families, respondents rated their families high in closeness.[11] They especially noted feelings of mutual concern and support. We found similar results in a study of adolescent girls raised in dual-career families.[12] The benefits frequently mentioned in both studies included children's perceptions of both parents as positive role models, their exposure to a broad range of values and experiences, the family's financial security, and the children's opportunity to develop independence; time constraints were the most frequently mentioned problem.

Several girls commented on the positive influence of parents, particularly mothers: "The decisions mom made seem like the best choices, the way I'd decide on my own probably." "My parents have shown me that a female should not have to take care of the house alone and work. There should be equal sharing of housework." A common theme was that "they want to make sure I can live on my own." Many of their parents emphasized the importance of a good education, responsibility, and setting goals to achieve.

My own daughter said she sometimes feels we don't have enough time for her or get too distracted by pressures at work. On these occasions she rates growing up in a dual-career family as a 4.5 (on a scale that ranges from not that bad, 1, to terrible, 10). Otherwise she rather enjoys it. When I asked her what she saw as the effects on her, the immediate response was "getting to go to a lot of places, knowing a lot of different people, and going to your and dad's offices." This was followed by learning good habits, the ability to communicate (especially with her children, should she ever have any), and the willpower to stick with things even when the going gets rough.

I think her last point about willpower relates to experiences with sexism. She was distressed by a recurring situation at school. The boys assumed they were smarter and more skilled than the girls, even though evidence made the contrary seem obvious. Not even logic could alter their viewpoint. The biggest insult occurred in physical education, which was a coeducational class. She was by far the best soccer player in the class, but even her male teacher failed to recognize her ability. She sat out a good part of the class time and watched the boys play. It infuriated her (and us).

What can parents do when they encounter prejudice? Most important is making sure the child recognizes the prejudice and does not see herself or himself as lacking in some way. In our situation, we assured our daughter that she had excellent soccer skills and reminded her of the evidence—numerous trophies won by her team in the city's youth league for girls. The next step is pointing out the prejudice to teachers, coaches, principals, or whoever is involved. Most people don't want to hear it, but they may eventually listen if parents remain persistent and nonjudgmental. It's a matter of education. Few people these days choose to be sexist with children.

Many children from dual-career families become quite articulate and assertive about sexism and develop a sensitivity and awareness about sex-role issues. The comment of a son in his early 20s seems particularly poignant. He believes that the main advantage to him was seeing his mother, and therefore other women, as his equals. Witnessing parents' struggles with taking on nontraditional tasks helps children comprehend the power of sex roles. When mothers encourage and reinforce fathers' participation in traditionally feminine household chores, which is often the case, their understanding of how men are changing heightens. Moreover, a more egalitarian-type marital relationship gets modeled.[13]

This leads us to a final question: How "nonsexist" is the parenting in dual-career families? Bringing up a child in a nonsexist way can be a subtle and complex process. It entails providing certain experiences and a great deal of support to broaden traditional views of maleness and femaleness in a society that is still quite conservative. A recent headline in the *Wall Street Journal* read: "It's Business as Usual at Toy Fair: Familiar Themes and a Harder Sell—Violence for Boys, Love for Girls."[14]

Parents interested in raising their child in a less sexist manner report particular frustration in locating caretakers who share their philosophy. Schools and teachers often reinforce and perpetuate sex–role stereotypes. In the end parents have to compromise, placing their children in schools they do not really like and with teachers who have traditional views about gender. Many examples come to mind. One day my daughter came home absolutely delighted with herself. She had "wowed" directors during an audition with her impromptu rendition of a girl dealing with a sexist soccer coach. She said, "It was so easy. We have been through this hundreds of times." On the less humorous side is a situation that infuriated the parents of a 12-year-old girl who excelled in math. She was asked to tutor other children in the class while the teacher provided the boys who excelled with advanced instruction. Parents of sons get irritated by the pressure put on boys to choose athletics over drama and music. One boy, very talented in ballet, gave it up because the other kids thought he was weird.

Nonsexist child-rearing at this point has different implications for the upbringing of daughters and sons. Women in dual-career families pursued nontraditional paths and know the importance of instilling similar values in their daughters. Many parents in dual-career families, mothers in particular, want their daughters to grow up with both the proverbial "roots" and "wings."[15] Roots, which girls traditionally developed, provide a sense of commitment and dependability and an ability to connect to and nurture others. Wings, traditionally reserved for boys, give them the strength to be independent, take care of themselves, and evince a sense of well-being separate from others. One father noted the changing times and the excitement he felt watching his daughter mature from a girl to an adolescent[16]: "This little girl of mine is tough, with creative aspirations that challenge her to write ballets about swans while shedding sudden tears when a friend moves away. I look forward to watching her grow and mature." Toughness and tenderness fit for girls who grow up to be independent women who know how to nurture others.

But double standards still bend our perspectives, especially when it comes to sons. We still live in a patriarchal society in which men hold most positions of power and leadership. Overall, men's lives have changed less dramatically in the past 25 years than women's, even for many partners in dual-career families. Because society has changed less for men, less attention gets paid to bringing them up in a nonsexist way. Moreover, parents may feel conflicted about the fate of sons who develop softer, more feminine qualities. Will sensitive men survive as well? Will feminine qualities make sons less powerful and successful in the long run?

Although their motivations may vary, parents feel greater social power to maintain the traditional qualities in their sons. As a result, parents intervene less in sons' choices and provide less encouragement for sons to step out of the prescribed "male" line. Few parents would question a 14-year-old boy's choice of shop and judiciously mention the importance of home economics or typing, even if the latter better fit his interests. Rather, most would take the view that "boys will be boys" and refrain from actively promoting values and behaviors that fall in the stereotypic feminine area. I was recently at a baby shower for a colleague to which people brought their young children, ages two to six. At one point the children got out of hand and were jumping on the furniture. The daughters were immediately disciplined, but not the sons. One parent said, "Boys just do that. You just can't keep them under control." People standing around nodded in silent agreement.

This tendency to "let boys be boys" seems shortsighted. If anything, at this point in our society nonsexist rearing of boys is probably more important than—or as important as—rearing girls in a nonsexist way. Otherwise, the individual struggles will simply be repeated with each generation and larger social changes will be minimal.

* * *

Parenting in dual-career families remains complex. Women have pushed for dual-career marriages, and they experience more stress and overload than men when it comes to caring for children. Affordable child care persists as a trying and burdensome problem. Yet most women see the situation as positive and would not change it for a career in full-time homemaking. Men are being nudged by the women they love to participate more actively in parenting, and they too see the rewards in terms of their expanded roles within the family. Such changes cannot occur without conflict, however. Partners need to know what problems to expect and how to effectively manage them.

CHAPTER 6

TWO CAREERS IN THE SAME FAMILY: THE CAREER SIDE

> Would anybody ever write: "John Jones is both an entrepreneur and a father"?

Career professionals expect some kind of reward for their hard work. This varies among individuals but typically takes the form of status, money, autonomy, or a sense of having made an impact or difference. Each partner in the dual-career couple has a particular definition of success, which often includes a personal or societally imposed timetable for where each should be when. But women and men choosing careers today do so from vastly dissimilar historical and sociological perspectives. Men have always pursued careers. In contrast, for women, having a professional life represents a recent and dramatic broadening of options for personal fulfillment.

Because "his career" and "her career" emerge from such different societal norms and expectations, partners' interpretations of what a career "means" likely differ. What one partner, say the husband, assumes is true of careers, the other partner, or wife in this case, may find unnecessarily restrictive. For instance, many career men believe professional demands must be met regardless of the effect on relationships and families. Many career women, in contrast, want to combine families with their professional lives. A second example concerns the assumption that careers are undertaken on a *full-time basis* as a life work. From a broader perspective, a spouse can be committed to an occupation and have an employment pattern that does not meet such a strict definition; part-time work or other adaptions need not reflect a lack of employment commitment or of the importance of a career to one's self-concept. As society is presently organized, however, most women and men committed to pursuing occupational careers must do so full-time in order to be successfully employed and to progress in their fields.

Moreover, what is standard procedure to one partner, say the husband, is to the other an outgrowth of male dominance. Maternity leaves provide a relevant example. Women, not men, typically receive this benefit because parenting falls within women's traditional purview. Such a discriminatory practice, however, helps maintain male dominance in the workplace; the man's professional life continues undisrupted.

One spouse's evaluation of successes and failures may also bear little resemblance to the other's. Many men know that they belong at work and readily structure their work lives as if they do not have families. These assumptions carry no agony or conflict and are free of anticipated risk or sacrifice. Moreover, families generally foster men's work goals by providing support, encouragement, and stability. And personal success makes men more desirable—to employers, to women, and to families. Personal success is what it's all about. Women, on the other hand, traditionally remain at home or in jobs that readily accommodate family demands. Caring for families and assigning men and children first priority are the givens; personal success is neither anticipated nor applauded.

But women who pursue careers, not unlike men, must structure their professional lives as if they do not have families. What results is somewhat of a paradox: The structure of careers assumes people do not have families and the structure of society assumes women care for families. Women who desire careers, then, need to create a new reality, which in turn may generate conflict, guilt, and self-questioning. The emergence of the dual-career family lifestyle represents this new reality. It provides a unique and creative solution to combining career and family. However, success at this emerging lifestyle invites a host of new stresses, ranging from husbands' resistances and anxiety to the necessity for wives to spend more time away from home.

Thus, "her career" in two-career couples, not "his career," rivets the attention of social scientists and media experts, although few probe beyond the surface of the woman's dilemma. Study after study discusses how professional women cope, why they should or should not be in the board room, how their careers are affecting their spouses and children. Magazines tell working women how to combine their roles. And a new form of female fashion for the emerging executive women is designed to fit women into the world of work and to project an aura of competence, confidence, and sexual and personal power.

The scale is out of balance in the men's direction as well. Little recognition is given to the dual roles (work and family) of married professional men—even for those living in dual-career families. My daughter was so surprised that an article about her best friend's father made no mention of his family! This would rarely occur in interviews of married professional women. The world wants to know if career women have families. That's still new and different *and* hard. Men having careers and families is old hat and supposedly not all that difficult, as the introductory comment about a fictitious John Jones indicates. Clearly, society's views of men as dual-career partners need to be revised and en-

Doonesbury BY GARRY TRUDEAU

lightened. Men need to be given more consideration as equals in their own way—as this "Doonesbury" strip suggests.[1]

His and Her Careers

> To make a dual-career marriage work you have to sacrifice a little for each of the careers. [A husband in management whose spouse is in sales]

What happens when women do not channel their talents and energies into their husbands' careers and the couples are without traditional "wives"? Some argue that spouses in dual-career families just can't compete with professionals who don't attempt to combine their careers with active participation in domestic activities. A person's career supposedly takes priority over all life involvements—including family. That's the way it has always been and that's the way it is now. If some couples wish to be innovators and integrate work and family, they will have to forsake careers that require substantial training and commitment for nonprofessional jobs that lend themselves more easily to interruption or part-time employment.

Others argue that careers are not the enemy of parenthood. The enemy is the way people think about careers. Because careers are supposed to be totally fulfilling, individuals either let careers expand into all the available space, worry when they do not, or assume that careers free them from other life roles. Adversaries of the "all-fulfilling career theory" see such reasoning as status quo thinking. It benefits those in power, especially men who aspire to or have professional careers, and the existing societal structure. In reality, most people's careers are not all that high-powered and all-consuming. Physicans, psychologists, and lawyers, for example, may put in long hours, be devoted to their work, and conscientiously keep up with their fields, but their careers are rarely out of the ordinary or all that engrossing, and this fits with their expectations. Success is measured not by achievement *per se,* but by money—hourly fees, annual incomes, commissions, and so on. A sizable number of professionals work 35-hour weeks. Those who work longer hours often do so to earn more money, a motivation that may be lowered by having a spouse who also earns a reasonable salary. Proponents of this emerging modern perspective, then, view dual-career families as a viable alternative.

How the Two Careers Progress

> I changed from a high-pressured job that required a lot of travel and weekend work to my present position [owns his own small business] so I could spend more time with my wife [a physician]. I could start my own business because of her salary. [A man in his 30s who has a preschool-aged child]

Being in a dual-career family influences the careers of most spouses, male and female, although few label the effect as negative. Rather, most perceive the influence as positive or neutral. Many men feel their career has been enriched and their opportunities enhanced.

They worry less about money and feel more freedom to explore other career options. Women, too, perceive family life as adding to the rewards of their career opportunities.

There comes to mind a conversation that took place some years ago when my husband and I were newlyweds. A famous scientist in a two-career family, who later won the Nobel prize, advised us about the importance of family to successful careers. He himself used his time with care and efficiency and rarely worked evenings, not only because he was a morning person but also because evenings were for family life. Family revitalized him for the next hard day. His advice well reflects the sentiments of many spouses in dual-career families.

Many couples see assets in having a career spouse. Foremost among these is having someone who understands career struggles and can provide emotional support when things get tough. One woman, who was overlooked for a promotion and finally decided to move to another company, said: "There are high times and low times. When things are going well, I take success for granted. When they get tough, I remember the benefits of a seasoned career spouse." Spouses who are highly ambitious may reinforce each other's work habits. Such couples may decide not to have a child—a decision that depends on the circumstances and that may or may not be viewed as a cost to the individuals involved or to society.

Other assets include added prestige from the partner's career and access to professional expertise and contacts in another area. A father in a role-sharing marriage explains: "My colleagues are impressed that I am married to such a successful and delightful woman. Her contacts have proved invaluable to me in my work and her zest for her work inspires me in my work. Sometimes I think they are envious of our lifestyle although they are not willing to consider a dual-career marriage for themselves."

The biggest tradeoff in terms of careers is less time to devote to work, which, in turn, may jeopardize career advancement and personal satisfaction. Whether it in fact does is impossible to determine. When I asked spouses if they would have accomplished more had their career been the only one in the family, many couples pointed out that it was "a moot point—two careers was a condition of our marriage." Spouses typically don't think in these terms. Rather, they focus on how much more their lifestyle allows for each of the partners. Often there is a sense of pride in each other's ability to accommodate two careers with family life and to willingly do battle with the traditional bastions of male dominance and female nurturance.

A midlevel manager in a role-sharing marriage adds another dimension. He supposed there was a substantial cost to him careerwise because his commitment to his daughter is not accepted in the business world, but that's not how he viewed it: "My relationship with my daughter is paramount. It can't be measured in career terms."

Little evidence supports the dire prediction that two careerists in the same family are doomed to uninspiring careers. In marriages with no children, the careers of both spouses progress more or less in parallel and on a par with those of peers. In marriages with children, husbands traditionally advance more quickly in their careers than do wives. The greater the wives' domestic responsibilities within the marriage, the greater the impact on their career advancement. Husbands in role-sharing marriages report more of an influence on career choices and directions, but overall, for men, the effects appear benign.[2]

Most dual-career couples are highly committed to this lifestyle and do what it takes to make it work within the constraints of their situation and the expectations for their career. Couples who succeed at this lifestyle see it as maximizing possibilities for both spouses. The whole is so much more than the sum of the parts. If one spouse is more career-oriented, the other, whether wife or husband, takes on more of the home load. If both value family involvement, they do what they can to achieve this. If both spouses put careers before everything, children will not be a consideration.

Who Earns What—The Question of Salary

> It doesn't bother me that my wife makes more than I do. Does it bother you that she also makes more than you do? [A friend's standard response to inquiries about his wife's high-level position]

What about salaries? The relative incomes of spouses can have significant psychological and practical ramifications in a marriage. Men expect to earn more than women. That's another part of their good-provider role. If they do not, they may feel less successful, which could have negative consequences for the marital relationship. A highly competitive man, for example, would struggle with earning less than his spouse, particularly if his self-esteem were closely tied to external manifestations of success such as salary and prestige.

Also, the one who brings in more money often has more say about how it is spent and less responsibility around the house. This could be especially troublesome if the wife is the more ambitious and higher paid of the couple and the husband feels reluctant to do more than his "equal" share of the family work. Establishing a sense of equity in the relationship would be difficult. More than that, there may be the subtle guilt of the spouse who makes less and has to spend the other's hard-earned money. A woman attending a recent lecture I gave on understanding dual-career marriages asked if I had studied "the problem of men who become less ambitious after marriage and are content to have their wives earn more." Some see this as unacceptable and ultimately unfair to the wife. But why should husbands be more ambitious than wives? Ambition varies in men, just as it does in women. The crucial issues, as I discussed previously, center around whether he does a fair share of the family and household work and whether the couple feels comfortable with the situation.

Most men earn more than their spouses, however. This remains true for spouses in dual-career families, although to a lesser extent than for couples in general. Recent reports conclude that "in wages, the sexes may be forever unequal"; women now earn 68 cents for every dollar a man earns.[3] (Not too long ago the figure was 59 cents.) The wage gap is narrowest today for younger women and for women who have had the same education as men, chosen the same fields as men, and lived the same work lives. The gap is wider for women who have gone into traditional fields and for women who have taken time out for childbearing and child-rearing.

Much ado is made about women who earn more than men. But less than 10–15% of wives in dual-career families out-earn husbands. How these husbands respond mostly depends on the stability of their self-image and masculinity, on their satisfaction with their own career advancement, and on how ambitious they are. Several men admit it might bother them if their wife advanced more rapidly and made more: "Intellectually that's ridiculous, but it would." Many others relish the thought: "In the end, greed overcomes tradition. I think I could learn to live with it." If husbands feel satisfied with their careers and mutually supported in the marital relationship, differences in salary affect them very little. Problems are more likely to emerge when the spouse's salary difference becomes linked to expectations for domestic responsibilities that the husband finds inequitable or to disappointments about the time available for the marital relationship and family life. In later chapters I discuss the key role of "time for each other" in developing marital quality.

Typically, partners report little discomfort about husbands out-earning wives. This is especially true in traditional dual-career marriages in which the husband earns a good deal more and the wife takes on more at home. Problems can come up, though, in even these marriages. A common example is when the husband earns more because he has a higher degree or is in a more prestigious field than his wife, but she works harder and is more ambitious careerwise. He goes skiing or plays tennis on afternoons off and she rushes home to start dinner, on the way picking up the children from afterschool care. For the husband, having a career gives him the privilege to pursue personal interests in his free time. This upsets his wife because she sees herself as the more career-committed of the two and as doing more than her fair share. Working through this dilemma requires a significant attitude change on the part of husbands. In fact, this may well represent a situation that requires professional counseling.

Discomfort can emerge in ambivalent and role-sharing couples when "earning more" gets translated into "doing less at home." Some husbands feel resentful when wives work long hours for relatively low pay and expect the husbands to share home responsibilities. The husbands' attitudes in turn infuriate their wives. Effectively handling such situations requires wives to confront husbands' self-serving views and husbands to reevaluate the traditional index of masculine success—namely, money. Female-dominated professions such as nursing and teaching offer lower salaries. Less importance should not be attributed to wives' careers simply because their positions pay less. Negotiating this situation can be tricky, however, because wives, too, may view female-dominated careers as less worthy and, as a result, abdicate their initial position that husbands should do their fair share.

Sexual Discrimination and Prejudice Still Make a Difference

A female vice-president of a bank feels frustrated and dissatisfied at times:

> I know I have the capability to be a part of senior management. If I were a man, I would have been there long ago.

Her husband, a landscape architect, is equally annoyed:

> My wife should be promoted and make more money. She performs as well as the men and this should be recognized and rewarded.

The structure of the labor force continues to show marked discrimination on the basis of sex. Women often have difficulty entering male-dominated fields such as business and engineering, and those few women who are admitted typically receive fewer promotions than men with similar credentials and are directed into career paths that are lower in status, power, and salary. As I mentioned earlier, despite the influence of the feminist movement, women at all educational levels continue to earn considerably less than men and are underrepresented in nearly all professional areas. Only a small percentage of employed women hold high-level executive, administrative, or managerial positions, and women occupy only about 3% of the 16,000 seats on the boards of the thousand largest corporations listed by *Fortune* magazine.[4] Women's acceptance and promotion in careers continue to be based on factors related to their sex rather than to their ability.

Moreover, employers may be skeptical about hiring women who are attempting to balance professional and family demands (or men who participate heavily in family work). I know of a woman recently promoted to middle management whose boss likes to call breakfast meetings. This proves terribly inconvenient to her family, but she is reluctant to say anything. She knows it will be viewed negatively. The message is clear, she said: "Work comes first. They act as if no one has a family." These breakfast meetings would likely affect male executives in dual-career families in a similar fashion.

The stress of prejudice and unequal opportunity for women affects men too. The husband mentioned in the introductory quote to this section was terribly upset by his wife's situation. Not unlike many spouses, he assumed sexual discrimination and harassment was part of the past—until it happened in their lives. A man who shifted careers and now shares an accounting business with his spouse could not believe how often "this male chauvinist thing with customers raised its ugly head." Customers wanted to work with him, not her, although she was clearly the more experienced and knowledgeable.

Experiences such as these open up men's eyes to what women often endure in their careers and, as a result, modify their own behavior with women. I know several husbands who, after having their consciousness raised by their wives' experiences, brought about dramatic changes in their own work environments. Special workshops were arranged for managers in two cases. In another, female staff at a professional agency were asked to monitor and then summarize in a report instances of prejudice and harassment. The report was then discussed at a staff retreat.

The *Wall Street Journal* and the Gallup Organization jointly surveyed 722 female executives with the title of vice president, or higher, to find out how it feels to be a woman in the corporate world.[5] The disadvantage most frequently mentioned by the women was the prevalence of male chauvinism. Many commented that a woman is still not taken seriously at higher levels. For example, 70% felt they were being paid less than a man of equal ability; 60% said their views were not respected as much as a man's. Also noted by many was social exclusion from the male turf. The flow of business information between men is informal, but between women and men it is formal. Men get together for lunch or drinks after work; they chat in the halls or on the way to play squash. In contrast, conversations with women typically take place at the office and at appointed times. Sexual harassment is also a fact of life even for women who own their own businesses.[6] When there are no bosses, it comes from clients, suppliers, bankers—indeed, from among all groups of men with whom the women deal.

Sometimes women are passed over for jobs or promotions for reasons that have nothing to do with their qualifications. A good friend feels powerless. His wife was in line for a promotion but it was offered to a male, new to the company and ten years younger. She was crushed and demoralized. After many discussions, they decided against legal action, but she still needs to face this situation every day. Coping with the situation proves difficult, however, and she plans to leave as soon as she locates a comparable opportunity. Another possibility is a career shift, which she had been toying with before this setback occurred.

Husbands can also inadvertently influence whether companies hire women. The burden of having to transfer a husband (or a wife) may be sufficient reason to pass up a woman for a promotion requiring relocation. Also, although rules against doing so are less prevalent these days, many companies decide against hiring a husband *and* a wife.

Women pursuing careers still buck a traditional norm and as a result face counterpressures such as smaller salary increases and fewer promotions. Husbands can provide a trustworthy "reality check" in these instances and help wives develop effective strategies when needed. For instance, couples can role play how to approach a boss about perceived inequities in salary increases. Husbands can also help wives deal with stereotypic and at times hostile responses from other men. Wives may be labeled too aggressive and pushy or too passive and gracious to function effectively in their careers. Women who return to work shortly after childbirth may be viewed as "unfeeling." Wives must view these kinds of events objectively, not personally, and husbands' perceptions can facilitate their doing so.

Not all women experience sexual prejudice or harassment. And some companies recognize the need for education in these areas and regularly sponsor seminars and training programs on sexual harassment, sex bias, and male–female collegial relationships. All the same, couples need to know that sexism is alive and well in the business world. Their best defense is knowledge. Couples need to be prepared to recognize prejudice when it occurs and use the necessary legal options and company-based policies to counter it.[7]

Locating and Relocating—Finding Two Positions in the Same Place

> I wouldn't mind following her for her career. She has as much of a right to work elsewhere as I do, but I feel there are relatively few places that I would consider. [A man in his late 40s married to a woman he views as much more ambitious and occupationally committed than himself]

Finding two equally attractive job offers within reasonable geographic proximity in a desired locale is no small feat and indeed may be impossible. In fact, finding a job of choice that requires moving from a current locale may very well be the most difficult issue for members of dual-career families. One or both spouses may fear a decrease in career options. As one man said, "What I do now is tempered by what my wife does." Also, spouses may have to settle for something less than they want or feel they deserve. Nearly all partners struggle with this at times. One very well might be happier professionally elsewhere, but moving, given present family circumstances, could bring more costs than benefits. So the couple stays put. Most couples typically can absorb low levels of dissatisfaction quite well, particularly if the more satisfied spouse shows sensitivity to the other spouse's situation. The impetus to relocate comes from one or both partners expressing considerable frustration and anger with their situation and feeling they can no longer "make the best of things," or from one or both receiving an unusual opportunity.

The experience of one role-sharing couple highlights several of the main issues for partners trying to locate two positions. Both spouses in this example finished their schooling at approximately the same time. Because his field was extremely limited, the couple agreed that he should find a suitable position first, and then she would relocate with him. After nearly a year of searching, the husband received an excellent offer from a university in a small Midwestern town. He had some reservations about taking the position, but after a good deal of discussion they both decided in favor of his accepting the offer. When they visited the town a short time later to look for housing, however, the wife discovered her career prospects would be dismal. She began to have serious second thoughts and felt fearful and angry that her life was being controlled by the job market in her husband's field. The possibility of his going alone came up, but they worried about how their two young children would interpret and cope with the absence of a father to whom they were very attached. The husband finally decided to resign the position and explained his wife's professional dilemma. A position he had been offered but turned down, which was in a larger city, was still open and he accepted it on a temporary basis. "So there it is," he observes. "Frankly we are not sure what happens from here."

This couple, like most two-career couples, wished to give equal weight to the interests of both partners. The reality, however, is that women and men often relocate for different reasons. As this wife originally intended to do, women move more frequently because of spouses' relocation, whereas men, not unlike the husband in the example, relocate for a good position or in order to gain a higher income or greater job prestige. Locations or relocations based on the husbands' needs are still the norm. Husbands are usually older than wives and have more years in their professions. Also, men still generally command higher salaries than women. And individual income is by far the most critical factor in determining location or relocation. The more wives earn—both in absolute terms and relative to their husbands' incomes—the greater influence they have, and the more they expect husbands to take their careers into account, when contemplating a move.

The experience of this couple also reflects other important facets of the location–relocation issue—communication, honesty, and compromise. The husband in this example understood the wife's dilemma. He knew he could not be happy in a professional position when the prospects for his partner were nearly nonexistent. On her part, the wife wanted to provide maximum support to her husband, but after the visit she realized she had gone beyond what she could truly live with comfortably. She had gone beyond what felt equitable.

Although this was not the case in the example, men's own sense of entitlement may come to the fore when decisions about relocation must be made. As a different man remarked[8]: "It is difficult for the man in the family—me—to put my career at the mercy of my wife's. In the starkest psychological terms, I would be following her and abdicating my role as a man." Another husband, who left his university position so that his wife could seize a career opportunity with her company, described his mixture of pride and anger. He felt genuine pleasure at her success; "And yet," he said, "I was unable to quell my resentment at her. I felt that because of her, nobody knew who I was."

Most people with careers do not relocate. Moving is stressful in the best of circumstances, and all things being equal, individuals prefer to stay in their present situations because of the climate, their children, their family roots, or their positions.[9] Crucial to many, however, is the freedom to relocate should the opportunity present itself. The spouse who said that "what I do is tempered by what my wife does" is happy where he is, but he also knows that should his situation change, moving would be difficult. At times this freedom can take the form of a commuting marriage.

I recently saw a cartoon with the caption, "And they both lived happily ever after—she in New York and he in L.A."[10] Although this hardly represents the norm, women who relocate for their own careers often enter into a commuting marriage. Fewer men report such experiences; if they relocate for job reasons, their spouse usually goes along and there are relatively few negative repercussions. This difference again speaks to the fact that men have higher salaries. Both spouses may consider moving to accommodate the wife's career as too costly and choose to commute instead; in contrast, moving to accommodate the husband's career makes sense.

Few couples commute as a lifestyle; most live apart for relatively short periods of time—a few months to a few years—with the hope that both spouses will be able to locate or relocate in the same place. Commuter marriages are not for everyone, but they can be worth it for some. For example, a commuter lifestyle would likely be unsuccessful for couples who communicate poorly or who have little commitment to the marriage. Benefits of commuting include allowing time for one spouse to get established before the other relocates or time for the spouse who most wants to relocate to "try out" the position before the entire family moves. Sometimes one spouse waits until a son or daughter finishes his or her last year of high school before joining a spouse who relocated.

Finally, there's additional *good* news. The recent increase in dual-career couples has heightened employee resistance to moves, lessened the career stigmas formerly associated with turning down transfers, and prompted some companies, such as General Electric, to require fewer moves for advancement. The trend has also sparked a new benefit—relocation aid for the other spouse. Some 75 large corporations now provide the accompanying spouse with job assistance. But the help tends to be informal and limited, such as providing job leads and paying employment-agency fees. Some companies find positions within the company for the other spouse, but this is still quite rare.

Couples cope with career choices and the decisions they entail in varying ways. At the same time, one gets a clear impression from couples that some flexibility and sense of freedom of choice must always be maintained. Their sentiment is aptly captured by one person's comment: "We will stay here as long as we both can grow, but we will not stay here 'at all costs.'"

PART III

HOW TO MORE THAN MANAGE

Spouses in dual-career marriages deviate from the common rule—both the wife and the husband intend to pursue the *same* life role. With neither spouse designated to provide the traditional TLC (tender loving care), both spouses need to sustain the relationship. Thus, couples in successful dual-career marriages must learn to strengthen each other and to *more* than manage.

CHAPTER 7

STRESS AND COPING IN TWO-CAREER FAMILIES

> We are both determined and take the time to make our marriage work. We love each other very much and we are a good fit. Our roles are not real strongly defined. Among some of our friends the husband expected the wife to work all day and come home and serve him coffee. These couples are now divorced. [A role-sharing couple in their late 40s]

Dual-career couples do not have a corner on stress. Stress is synonymous with living and in itself is neither bad nor good. We all confront stress most of our lives. Eustress, or good stress, invigorates. Distress, or bad stress, debilitates. Stress that comes from challenging work, for example, often enhances creativity and brings more meaning to life. In contrast, stress that comes from stultifying work, thwarted goals, or the unavailability of suitable child care diminishes feelings of well-being and lowers life satisfaction.

What one person finds stressful another may not. Individuals vary in personal styles. Some people are the "anxious type" and worry about almost everything. Others appear much more laid back and take things as they come. Typically, individuals have certain spheres that especially cause anxiety, worry, and eventual stress; these typically coincide with areas of low confidence or uncertainty. Examples range from aspects of job performance to parenting and child-rearing abilities.

Individuals also view situations differently and attribute different meaning or significance to what occurs. Typically, the more important an event to a person's life, the better candidate it becomes for stress. If some event has little or no relevance to a person's well-being, the assessment of the situation takes on an essentially neutral tone. Little needs to be done to cope with the event and life remains unaffected. Similarly, if the demand or event can be handled with available resources, and promises only beneficial consequences, the appraisal becomes positive. Stress enters in when the evaluation of the situation signals a possible loss, threat, or challenge because present resources might be taxed or exceeded.

The following list of a husband's possible responses to his wife's receiving a job offer in another city illustrates the hypothetical range:

1. A husband who knows his wife has no interest in relocating would view the news as *neutral* or irrelevant to his career. The most mutually beneficial response on his part would be to express pleasure for her and to congratulate her on receiving such a fine offer.

2. Should the husband in (1) above feel competitive with his wife, or feel jealous of her professional recognition, his response would likely be *negative*. He might well deride the offer made to her, taking the attitude that it can't be allowed to affect his career, even though she has little interest in moving.

3. A husband who has been making plans to move to that particular city for some time would see it as *positive* or salutary. Now he can go ahead with his own plans. His response might be, "Wonderful. Terrific. Now we can go on with our lives."

4. A husband who has little intention of moving but knows his wife wants to relocate would view the situation as *stressful*, and in a *negative* way. That is, he does not want to move, sees giving up his position as a personal loss, and feels anger, fear, or resentment for being put in this position. His response might be dismay. He wishes this were not happening to him and he wonders how the relationship can survive.

5. A husband who is not opposed to moving and knows his wife wants to relocate would also view the situation as potentially *stressful*, but in a

> *positive* way. In contrast to the husband who does not want to move, the stress for this husband may take the form of a challenge. He sees an unexpected opportunity for growth or personal gain that engenders excitement and eagerness. His response might be joy tempered by anxiety. He feels pleased for his wife and himself but worried about the unknowns of a new locale and the kind of life they will have there.

Of the various options included in this example, the fourth situation presents the most stress for the husband. In actuality, how the husband in this kind of a situation responds depends on a host of other factors. For example, if the husband had been under the impression that any relocating would be a joint enterprise, but had not been consulted about this particular job opportunity, he might feel anger and disappointment. Alternatively, if he and his wife had discussed possible commuting, his response might be quite different. He might still feel angry about the change in their day-to-day relationship but supportive of her decision to take the offer because it would enhance the marriage in the long run.

Essential to understanding stress is passion. What we care about, believe in, and commit to—our beliefs and values—give life purpose. For some, this takes the form of commitments to certain values and ideals, such as family life or living in harmony with nature. For others, specific life goals are crucial—having a child, being a successful trial lawyer, or having time to run five miles a day.

Commitments and values have a double edge,

however. They comprise the passion that sustains us and gives life meaning, and they cause havoc when things go awry. The areas of our passions persist as sources of possible stress because they are integral to the fabric of our lives and greatly influence our feelings of well-being.

A case in point for dual-career families is parenting. A child enhances life and represents an important life commitment. Should things go wrong with child care, either real or imagined, tension arises and disruption of other life goals and commitments may occur. Depending on the circumstances, not being able to find suitable child care, for example, can cause enormous stress for dual-career couples and may even result in a reevaluation of priorities. Spouses may be forced to change their game plan and take a longer leave, rearrange work schedules, or move to a less demanding position.

Less dramatic are the little self-doubts that can creep in when everything does not go smoothly or as anticipated. I always looked forward to picking my daughter up after school and hearing her excited, "Mommy, Mommy, Mommy." A few times, however, she complained about a game being interrupted or about not being ready to leave; I invariably experienced momentary pain. I would immediately think, she is not excited to see me; is something wrong? (Nothing ever was.)

Personal goals and ambitions also relate to passion and possible stress. We expect many things of ourselves but often fall short of the standards we set. My daughter struggles with the need to be first in her class. I point out to her the folly and pain in this, yet I know all too well where she is coming from. I, too, was brought up to be the best at what I do and I, too, confuse this with the idea that I am only doing *my* best when I am doing *the* best. Individuals in dual-career families seem especially susceptible to this struggle; often they are high achievers and hold themselves to standards that may not be realistic or healthy. Striving to be perfect as a parent or professionally invites stress.

Persistent feelings of guilt or annoyance provide obvious clues to expectations that require rethinking. Changes and often simplification in everyday routines can provide much-needed relief. One couple, irked by the mountains of momentarily fashionable but unironed cottons that were never available for wear, has decided to give away everything that needed ironing before it went out of style. Another couple, concerned about time away from their children, combines socializing with family activities. Their friends know that their two children are to be included in invitations for dinner, and they do the same. Entertaining has become a lot more informal and that's okay for now. Other couples limit time devoted to their careers. Evenings and weekends are reserved for family.

A common source of anguish, and one that I certainly experience at times, is a fear of giving in to the "patriarchal structure." Individuals sensitive to discrimination against women, for example, see things that many of their peers remain oblivious to. They recognize the barriers to women's career advancement and men's participation in the home. They chafe at traditional policies that serve to separate parent and child and ignore employees' responsibilities to family life. To some degree these women and men have to live by the rules in order to survive professionally and in their marriages. At the same time there is a gnawing feeling of selling out and being co-opted rather than striving for social action and change. You attend Saturday morning meetings or work late rather than say you just can't do it, that's family time. You stop pointing out examples of sexism to colleagues because it makes them more resistant, not more accepting.

Clearly, certain things need to be overlooked to function effectively and comfortably at work and in the culture at large. We keep a sign on our refrigerator to keep us in tune with this reality: "No one is going to give women power and influence because it's fair, right, or good." Some kind of moral compromising is necessary to minimize stress. Spouses need to carefully choose their ethical and moral battles; fighting every battle invites failure and greater upset.

Strategies for Coping

Coping strategies refer to overall plans of action for overcoming stress. Some stressors stem from self-imposed standards to perform better at work, to live up to expectations of being a good parent or spouse, and so on. Others come from demands imposed by others—either people, such as spouses, children, bosses, or parents; impersonal societal institutions, such as child-care facilities; or customs, such as how jobs are structured. Coping strategies become important to both short-term and ongoing conflicts. Thus, strategies help in making one-time decisions such as whether to take a different job, meet a deadline, or care for a sick child one's self.

They also assist in managing continuing stress related to the daily demands of work and family.

Most people's ways or strategies for coping fall into three general types:

- *Understanding:* Understanding the source of the stress
- *Management:* Managing the stress
- *Change:* Acting to change the source of the stress

A common situation will help illustrate these three ways of coping. The endless nature of career demands often causes ongoing stress for spouses in two-career families. Partners desire a balance between careers and family life, yet they could work 20 hours a day and still not feel caught up.

Let us assume a situation in which the demands of work are never-ending. There just isn't time to do all that seems required. In this situation:

- A spouse using *understanding* as a strategy would think: "No sense worrying about what I can't get done; I'll do what I can. Besides I like everything I'm doing."
- A spouse adopting *management* as a strategy would go in earlier and take a sack lunch.
- A spouse taking *change* as a strategy would start saying "No" more often to make his or her work load more manageable.

Understanding strategies thus involve modifications in attitudes, which in turn change the meaning that the situation causing stress has for the individual. In the example, the spouse shifts his or her thinking from "I have to get all this work done" to "I will do what I can." This reflects an underlying change in attitude about professional demands. Rather than assuming all requests should be met, and feeling stressed because there is no reasonable way to do so, the spouse restructures the situation cognitively and takes the view that doing the best she or he can is sufficient. Similarly, parents using understanding-type strategies for conflicts between professional and parental demands might think, "It could be a lot worse," or, "This is a natural reaction for working parents." This kind of strategy is particularly effective when individuals feel assured they are doing the right thing. They feel optimistic about the long run and see everyday hassles as relatively insignificant in comparison. They have little motivation to make actual changes in behavior.

Management strategies reflect attempts to "do it all." The spouse in the illustrative situation assumed all the work had to be done and decided to squeeze in more time by going in earlier and working during lunch. People who use management-type strategies try to meet all existing demands; they work more efficiently and plan their time more carefully with the idea of fitting in everything. Spouses who rely heavily on management strategies fall prey to the "superwoman" or "superman" syndrome described earlier. They expand their roles, making all the necessary accommodations. "Superwomen," the emerging stereotype of women in dual-career families, for example, add the new demands of a career onto the traditional careers of wife and mother. That time and energy are finite means, of course, that management strategies are effective only for transient or short-term problems. Reliance on such strategies as a lifestyle would be counterproductive. Burnout, poor health, or total exhaustion would be inevitable.

Change strategies, in contrast, attempt to modify the situation at the root of the stress. Thus, in the example, the spouse employing a change strategy looked for a way to change an apparently impossible situation. Rather than adapting to the situation and trying to do it all or thinking about the demands differently, the person altered his or her own behavior. By judiciously saying "No" to some requests, the pressure of too much to do could be decreased. People who arrange work schedules with an employer to allow time for certain parenting responsibilities, negotiate schedules with a spouse, or hire a part-time bookkeeper for family affairs use this kind of strategy.

Change strategies require modifications in how partners define their life roles. For example, a person may view family or career as coming first, alter career or parenting aspirations temporarily, or change personal standards for home or work-related activities. A good friend decided to work fewer hours until his child was older; for him, time with his son became more important than additional income from his private practice. This decision eased his ongoing stress about not spending his time the way he wanted to.

The table below provides two additional illustrations of the three kinds of strategies.

Most people develop a pattern that incorporates some combination of these strategies and then modify

Examples of Coping Strategies Used by Spouses

SITUATION: Because of afterschool activities your child cannot take the bus home and needs to be picked up at 4 P.M. most days. You and your spouse typically work till past that time.

Understanding: Take the attitude that your child is getting to do what he or she wants to after school and you need to work until at least 5 P.M.; your child can wait and you will get there when you can.

Management: Dash out of the office as soon as possible.

Change: Call the school district and inquire whether there is a bus for students who stay after regular school hours. If there is none, get together with other parents with the same conflict and put pressure on the school district to provide this needed service. In the meantime, set up a car pool or arrange for someone else to bring your child home.

SITUATION: Your spouse likes to entertain and have friends come by often. You find entertaining too time-consuming, given everything else in your life. Resentment builds up on both sides because he feels deprived and you feel misunderstood.

Understanding: This is important for my spouse. Our relationship will be better for it in the long run.

Management: Greet friends graciously when they drop in unexpectedly and hang in, bleary-eyed and stifling yawns, until the last one decides it's time to go.

Change: Feel free to go on with your plans even if friends stop by, or inform friends when they arrive that it will have to be an early evening.

these patterns with time. Typically, spouses use understanding and management strategies more than change strategies in the early stages of career and family. When situations are new, they feel more scary. One almost automatically falls back on old, familiar behaviors, which may be more traditional in nature. If this reaction persists, it establishes a pattern, which may or may not be modified later.

Career women, for instance, often absorb more of the load at home for fear of rocking the boat with somewhat resistant husbands. Many men enter dual-career marriages as "armchair liberals"; they are all for role-sharing until they themselves have to do it. Women, too, have internalized beliefs about their roles and hesitate to push. Thus, wives might initially use understanding-type strategies in interacting with a spouse about housework. Stereotypically, the husband drags his heels about cleaning; the wife copes by reminding herself about all his good qualities and excuses this one area. Essentially, she changes the meaning of the conflict from "He is a resistant husband" to "He is not a perfect husband." So both spouses may rationalize the lack of role-sharing, an understanding strategy. The wife, in the meantime, copes by means of a management strategy and takes on more and more of the household work.

When they are younger, spouses may lack the confidence and experience to push for what they need in their home life or work situation and absorb the demands with management ways of coping. The earlier years of child-rearing provide a good example. Parents tend to use management strategies in interactions with children. Giving in to children's demands seems easier at the time; parents are tired and don't need the hassle. Also, "doing it all" may temporarily assuage feelings of guilt about leaving a child at day care or inadequacy about not being the ideal parent.

The change strategies become more frequent and distinctive as individuals mature and know themselves better. A personal example is my daughter's requests for fresh supplies of homemade cookies and cake, which didn't always fit with my priorities—nutritionally or timewise. I felt bad when I said no and a few times awakened at 5 A.M. to bake what she wanted. I soon saw the folly in this management style of coping, and shifted instead to a change strategy—we got in the habit of baking together and now she enjoys baking by herself as well.

Age and experience bring greater self-confidence as a spouse and parent. Over time, for example, partners learn to become more mutually supportive. Concerns about their own personal and professional adequacy take on less importance and ways of maximizing both careers and family life take on more importance. Wives who initially expressed some reluctance to push for role-sharing with husbands increasingly adopt change strategies with regard to domestic activities. Husbands, in turn, show less resistance because they too feel more secure and confident with their spousal and parental roles.

Also, time in a career puts one in a relatively established and secure position and enables one to negotiate from strength for necessary changes at work. Parents who feel more established in careers more freely combine family and career. For instance, a 38-year-old engineer with an excellent track record as a research manager assumed his employer would be surprised by his requesting a one-month paternity leave. However, he felt assured his request would not jeopardize his position or standing with the company. Generally speaking, with time and maturity, partners in two-career families can anticipate less stressful lives due to their experiences in coping and to the greater latitude they feel in altering customs and societal practices that are out of step with their lifestyles.

Resources for Coping

Strategies for coping are practiced within a context. We do not respond or cope in a vacuum. Often the effectiveness of understanding, management, and change strategies depends on other resources available at the time. These other resources can reside within the individual, such as self-confidence and knowledge. Or they can be present in the environment, such as suitable child care and a satisfactory position. A case in point is the birth of a first child. This event can be stressful under the best of circumstances, regardless of the type of strategies used. But couples who have their first child when both partners have established careers typically report far less stress than couples the same age who have their first child when both partners are in the initial stages of demanding careers. The more established couples feel more secure in their work, a personal resource. Moreover, they see themselves in a better position to negotiate for changes at work that would better accommodate parenting responsibilities. Thus, their access to environmental resources would be greater.

A certain amount of stress is normal and to be expected, as I noted earlier. Not all stress requires special managing and understanding or needs to be eliminated. In fact, the personal and situational or environmental resources accessible to spouses often keep tension and anxiety within normal bounds.

The accompanying table summarizes these two general types of resources. Physical health, a personal

Resources for Managing Stress

Personal resources
Health
Education
Satisfaction with current position
Personality: flexibility, openness, self-confidence, et cetera
Values: importance of achievement, marriage, parenting, et cetera
Communication style
Ways of coping: understanding, management, change
Family and other environmental resources
Spouses' compatibility
Spouses' ability to communicate
Mutuality of spouses' support for one another's goals
Each spouse's attitudes toward roles of women and men
Importance each spouse gives to family and parenting
Views of colleagues and employers about dual-career couples
Corporate policies about hiring of spouses, child care, parental leaves
State and federal policies about child care, parental leaves
Availability of suitable child care

resource, is a prime example. When partners feel energetic and enjoy excellent health, they typically manage the ups and downs of home and career quite well. When energy wanes because of acute illnesses or chronic health problems, what was easily managed now may cause stress. Resources also play a crucial role in dealing with inevitable stresses, such as finding two positions in the same locale or obtaining suitable child care, and with unexpected stresses, such as a spouse's illness or loss of a job.

Personal resources refer to a person's fundamental characteristics—attitudes, values, and personality as well as educational level, credentials, income possibilities, material or financial assets, and physical health. Access to resources such as income, education, assets, and physical health connects rather obviously with stress resistance. Typically, adequate education and training and a position corresponding to that preparation provide important resources for either warding off or effectively dealing with stress. Conversely, if spouses cannot find suitable positions or report dissatisfaction with their work, personal resources get somewhat diminished and stress is not as easily prevented or dealt with.

But when spouses consider the personal resources important to the success of their dual-career marriage, they refer less to finances and career paths and more to personalities, values, compatibility, and ability to communicate. A commitment to wanting the marriage to work takes top priority. Marriages make it when the spouses have a great sense of investment in each other and in their children. Similar family backgrounds and shared values add to a couple's general compatibility. Values about how to rear children take on special importance. Because both spouses participate in parenting, radically different philosophies might bring about some disgruntlement and possible discord, which would then need to be worked out between the two parents. How stressful such a situation would be depends largely on the flexibility and good will in the relationship at the time. Also pertinent to child-rearing are the values partners hold about the rights and roles of women and men. For example, if nonsexist child-rearing practices are more important for one parent than for the other, some conflict could develop and would need to be discussed.

Spouses' ability to communicate assumes a central position among the various personal resources. Although personal styles at times determine willingness to bring up troublesome issues and talk out problems, couples must find ways to talk on a regular basis. Communication, which is paramount to the success of the marriage, requires cooperation, flexibility, tolerance, and an ability to accommodate. It also requires time, which is at a premium in dual-career marriages. But commitments take time. My relationship with my husband is at its best when we talk—as things come up. We used to "wait until we had the time" and discovered that the time never came around. We had to make the time. Most couples do.

Communication patterns often relate to another crucial resource—spousal support. Spousal support is the fundamental *environmental resource.* Spouses thrive best when each acknowledges and values the other's various life roles. In actuality this means support by the husband for the wife's career and support by the wife for the husband's participation in parenting and housework.

Typically, men find it easier to value the woman's career than to partake in housework and parenting, however. Men committed to role-sharing may find it embarrassing to admit that their wives earn more of the family income than they do, and their discomfort may get translated into resentment about doing the cooking. Their wives, in contrast, may feel reluctant to put their career needs ahead of their husbands' and find themselves resentful about not taking advantage of opportunities that come their way. Such struggles definitely influence the ease and generosity of spouse support and underscore the importance of good communication. Spouses must talk to each other about feeling resentment or embarrassment, to both maintain and increase their goodwill and feelings of mutuality.

A sense of fairness or equity about the balance of family and occupational roles achieved by each spouse remains closely tied to the extent of mutual spouse support. When each partner feels fairness prevails, mutual support peaks. When either partner feels taken advantage of, especially for long periods of time and despite many heated discussions, the marriage suffers.

The key to achieving mutual support resides in communication and discussion—both the willingness to engage in conversation and the ability to listen to one another. Take the situation in which a husband feels his wife's career is too all-consuming and leaves too little for their personal life together or a wife feels too constrained in her career because of limited mobility or family demands.

Considerable personal and marital distress occurs in both cases, partially because partners feel disregarded and unsupported. The husband in the first situation wants more of a personal relationship. The wife in the second situation wants more latitude for her work. Neither can resolve the difficulties alone—the difficulties evolve from the relationship and need to be treated from that perspective. Open, honest discussion would likely alleviate the immediate stress. Rarely are relationship problems all one-way. If one spouse is unhappy, so is the other. Continuing the dialogue, perhaps with a counselor, would bring other dissatisfactions and disappointments to light and allow possible alternatives to emerge. (Chapter 8 discusses these issues in detail.)

Nonetheless, spouses in dual-career marriages sometimes find themselves struggling with their own sex-role socialization. I noted earlier that the smaller the difference between the husband's and the wife's income, the greater the husband's participation in the family and home. Moreover, husbands report the least stress both when wives hold positive attitudes toward the husbands' careers and when husbands involve themselves in housework and parenting. The same holds true for wives. Hence, mutual support for family and work commitments minimizes stress by providing a strong underlying resource or foundation for the marriage. Husbands traditionally were not expected to support wives' careers or to take a major part in the daily family life, and many still drag their heels. As women reach a greater parity with men salarywise, we may see an increase in husbands' willingness to share more fully in domestic work.

Environmental resources extend beyond the family to friends, colleagues, organized groups, and social institutions. The availability of support beyond oneself and one's immediate family gives a person a sense of belonging and general well-being—with a place in the world. Such resources, however, remain largely unavailable to partners in dual-career marriages because of the inconsistency between their lifestyle and traditional views. Thus, spouses for the most part cope with life's stresses by relying on personal and family resources and by learning as they go. They look for colleagues who share their views and develop their own networks of support. The dual-career path isn't an easy one to follow, but the shared rewards prove worth the greater effort.

Although broad societal acceptance of dual-career marriages has yet to occur, some increase in environmental resources appears evident. Examples include more flexible work hours, rethinking of transfer and relocation policies by employers, and greater career opportunities for women. The most concentrated efforts appear in the area of supplemental child care.

Locating quality day care remains a persistent and recurring stressful task. Concerns about the general unavailability of quality child care is pervasive among couples. Although some corporations and communities provide some assistance, by and large individual families must secure suitable day care on their own. And when they successfully do so, they consider themselves fortunate. One woman, a physical therapist and mother of two, had lost three child-care providers, for various reasons, in two years. The third time she began to wonder whether the stress was worth the satisfaction of work and seriously considered resigning her position, something she really was loath to do. The situation fortunately resolved itself: "Luckily my employer opened an on-site center and I managed to get the very last opening because of a cancellation."

As this example indicates, promising developments are taking place with regard to corporate involvement in child care. More and more companies are instituting progressive policies that recognize the direct connection between provisions for child care, on the one hand, and employee morale and productivity and corporate profit, on the other. Nearly all organizations provide for maternity leaves of 30–90 days, but paternity leaves are far less common, and few men take them when they are available. Nationally, about 3000 employers furnish some sort of child-care support, including cash subsidies to help defray day-care costs and contracts with established day-care centers. Corporate-sponsored child care is also on the increase. A small but growing number of companies, approximately 150 companies in 1987, have day-care centers at or near their offices.[1]

The picture becomes less positive at the state and federal levels, although even here some progress is evident. The United States persists as one of the few major Western countries with no policy guaranteeing women or men the right to parental leave and job protection.

And Congress has yet to pass even a modest parental-leave bill. Moreover, 32 states have been forced to significantly cut their federally funded budgets for child care.

But the tide may be turning.[2] Several proposals are under consideration by Congress, including an infant-care plan that would provide 18 weeks of unpaid leave for parents. A more ambitious and far-reaching proposal with strong Senate sponsorship would set up an experimental program of full-service schools. Such schools would have one staff of teachers trained to instruct children in the customary academic subjects and a second staff of child development associates specifically trained to supervise children and provide child care. Finally, the Act for Better Child Care Services (the "ABC" bill), which was drafted by a coalition of child care providers and advocates, was introduced in November 1987 by Democratic Senator Christopher Dodd of Connecticut and Representative Dale Kildee of Michigan.[3] The bill would give grants to states for expanded child-care services to families. The key Republican bill, the Child Care Services Improvement Act, introduced in early 1988 by Senator Orrin Hatch of Utah and Representative Nancy Johnson of Connecticut, proposes to expand day-care programs and provide tax incentives to firms providing child-care services for employees. Clearly, dramatic changes at the federal level may be in the offing.

* * *

Some individuals enter into dual-career marriages unprepared. They expect others to applaud their lifestyle and to accommodate their constraints. This happens rarely. Rather, successful dual-career couples gear themselves to finding solutions and seeing problems as solvable or ones they just have to live with. They use coping resources; they don't try to do everything themselves. They cut down on social engagements to be with their children; they arrange schedules to maximize time with the family. The individuals who manage best are often married to spouses who understand the demands of careers because they have their own.

If a wife and a husband discover they differ vastly in values and expectations and show little mutual understanding and support, they usually divorce. Commitment to the marital relationship is essential. Also crucial is a willingness to struggle with the difficulties of integrating career and family despite societal pressures to conform to sex-typed roles and behaviors. Characteristically, spouses report relatively liberal attitudes about sex roles; they support each other's career efforts and value each other's strengths. When it comes to coping, they strive for strategies that reflect redefinition, compromise, realistic expectations, and commitment. As one couple noted: "Both of us are accommodating people. If an expectation is not met, that is not cause for walking out."

CHAPTER 8

HOW COUNSELING CAN HELP

Most people need professional help when things go wrong and all else fails. They are not sure what to do next or they do not like the consequences of the things they are doing. Spouses may feel ready to walk out of the marriage. Discussions may turn into arguments that never get resolved. Resentment sets in because one partner sees the other as insensitive to his or her career development or resistant to participating in domestic responsibilities. Expressing or receiving love is thwarted by antipathy and conflict.

Friends might show concern, yet realize the situation requires more than they can provide in the way of listening and emotional support. Typically, in fact, partners seek counseling because someone else makes the suggestion—a friend, a colleague, or a relative. This person either knows firsthand the potential usefulness of counseling or has a close acquaintance who found counseling helpful, and suggests it as a viable option for the partner, couple, or family.

Counseling or psychotherapy most benefits spouses who want to succeed in a dual-career marriage but find themselves floundering and unhappy. Through counseling they come to see their lives from a broader perspective and gain an appreciation for their struggles. Underlying feelings and attitudes related to problem areas emerge and become central to possible resolutions. For example, spouses' values about the prerogatives and power associated with being born male are a sensitive and potentially explosive area, but one often central to the conflicts many couples experience. Men, accustomed to having greater importance and significance than women, usually resist sharing power in relationships until they are forced to. Women, too, are reluctant to share certain rights and immunities provided by the traditional female role. Conflict comes about because one or both partners see advantages to themselves or to the relationship for things to be different and push for it. But pressing for change, regardless of how sweet the hoped-for reward, perforce engenders some resistance.

In the process of therapy some traditional men, for example, find they are willing to make the changes needed and come to see them as benefiting themselves and their relationship. Others, however, discover that to appreciably alter traditional sex-role behaviors and beliefs would be too costly to their career goals and life plans. Still others become cognizant of discrepancies between their attitudes and behavior but will continue to live this "double standard" until circumstances or their spouses push for change. As one client said, "I think it is entirely unfair that my wife does everything. But as long as she is willing, I won't complain."

The handling of household responsibilities and parenting frequently causes difficulties for couples. The husband of a couple I saw for counseling felt unappreciated by his spouse. He was preparing dinner three nights a week and doing what he considered to be his fair share of the family work. What emerged during the therapy was how his sense of male entitlement was getting in his way. He felt that he should be thanked for doing his share—after all, men typically have not done what he was doing. This attitude made his wife furious. Was she to be thanked for being a successful businesswoman because women typically have not done so? Even more upsetting to her was his implicit assumption that she did not need to be thanked for preparing dinner the other four nights, because that's "women's work".

Important aspects of the therapeutic process for the husband were becoming aware of his unconscious attitudes about male and female rights and responsibilities in the marriage and understanding how he and his spouse could feel the way they did. Crucial to this process was getting to the point where both spouses could nondefensively look at the situation from the perspective of the other partner without fear of abandoning their own positions or life experiences.

Problem Areas in Which Counseling Can Help

What motivates spouses to go to counseling differs, of course, from person to person and couple to couple. Nonetheless, certain life events or conflict areas more predictably bring partners in two-career families to counseling. The five conflicts I describe below reflect common themes that emerge in my counseling of dual-career couples: managing stress, the "match" between the wife's and husband's careers, struggles with role-sharing, deciding not to have a child, and accommodating aged parents. The problems and conflicts that brought these couples to counseling, and the resolutions they came to, illuminate patterns in the lives of many other two-career families.

Managing Stress

Stress is deceptive. It both excites and debilitates, and the fine line between the two at times gets blurred. Caught up by the reasonable demands of children or the deserved successes of career efforts, spouses or couples may become heedless of escalating stress, especially when things are going well. They inure themselves to the stress or simply assume it comes with the turf. They walk faster, skip meals, drive above the speed limit, use drugs or alcohol to relax, and so on.

Many counselors provide stress management programs that focus on defining and recognizing tension—before things get out of hand. The goal is to minimize or prevent chronic problems through awareness and subsequent behavior change. Fortunately, stress has obvious telltale signs—headaches, stomach pains, irritability, ragged and on-edge feelings. A key aspect of managing stress is the ability to honestly appraise one's day-to-day habits. Patterns of sleep, exercise, eating, and drinking must be looked at, as must emotional states such as one's usual levels of agitation and worry. Resources for coping must be scrutinized and typical coping strategies carefully evaluated. The bottom line, of course, is change.

What and how to change varies, but for most people relatively simple steps produce significant and positive results. Good nutrition, regular physical exercise, and instruction in self-assertion and time management head the list. I swim every day without fail and have done so for years. When things go wrong, I go for a long swim. My husband plays basketball and squash. We both mastered time management skills years ago and make lists to aid our memories. We have a large calendar in the kitchen and record all activities as they come up. At breakfast we go over the day's schedule and negotiate any last-minute changes. This system keeps us organized but, more importantly, lowers the chance of our getting overcommitted. Also, we know what's expected of each of us that day and can plan our work schedules accordingly or make alternate arrangements if needed.

Sheila and Bob provide a good example of how nice people can lose control of their lives. They care about the world, follow through on commitments, and generally act responsibly. They recycle, eat nutritiously, participate in civic organizations, and volunteer for school fund raisers. If you want a job done, and done right, you call Sheila and Bob.

Such admirable qualities may exacerbate their stress but are not at its root. The problem revolves around making use of time, setting priorities, and saying "No." Sheila and Bob are so busy juggling schedules and handling home and work responsibilities that they have no time for each other or themselves. At the time they sought counseling, the quality of life was just not very satisfactory for either of them.

Instruction in time management[1] made them see how they used their time and why. Priorities on time use were often set by someone else or set by them to please someone else. For example, Bob played squash three afternoons a week with a friend at a time convenient to his friend, but not to himself. However, he felt reluctant to negotiate for a time more mutually convenient and kept on this same schedule for months. Similarly, Sheila had difficulty turning down requests to head up various community projects. An effective leader and organizer, she was frequently beseeched by groups to take on the job. How could she say no?

Once they understood how time was used, Sheila and Bob struggled with priorities and planning. They set priorities first for individual commitments and then for *joint* time commitments. For example, time together was built into their schedule rather than left helter-skelter. Long-range (more vacations with the family) as well as short-range priorities (time to read the Sunday *New York Times*) were identified and listed.

The final phase of the counseling focused on learning skills in assertion.[2,3] Assertiveness, or having the courage of your convictions, requires not only establishing priorities but also saying "No" to low-priority demands on time. Learning this skill requires considerable time and practice for most people in this situation, and Sheila and Bob were not exceptions. Bob was easily frustrated when others did not easily respond to his newly acquired assertive behavior. For instance, when his squash partner said, "How come you played at this time for six months if it is *that* inconvenient?," Bob's response was to immediately back down and then later see himself as an "assertion failure." Within a few months, however, both Sheila and Bob felt in control of their time and their lives and counseling was terminated.

The Match between "His" and "Her" Career Expectations

Modern fairy tales would have both spouses obtaining the positions of their choice, advancing professionally at the pace each expects, and retiring happily ever after. If only that were the case. Instead, what looks like the perfect position turns out to be terribly disappointing. Firms close down or relocate. Departments disappear. Interests change. Work environments become stultifying. Wives return to careers. Midlife brings new perspectives.

Two kinds of matches affect marital satisfaction for dual-career couples: the match between career expectations and career reality for each partner and the match *between* partners. When partners weather the storms of unexpected or planned changes in positions and report relative comfort with their career accomplishments, there is unlikely to be any issue. Resentment develops when one spouse has satisfying work and the other feels he or she is in the wrong field or has limited opportunities in the present locale. Sylvia describes her attitude in the relationship with her husband, Sam: "We moved here because Sam had a good offer. I am not the workaholic Sam is, but I did expect to find a reasonable position with a good accounting firm. After looking for some time I finally opened my own office but I hate it. I feel more resentful every day and our relationship is suffering because of it."

This situation may call for relocation by both spouses. Another option entails career reevaluation, and perhaps even career change, for the dissatisfied partner. One benefit of the dual-career family is its flexibility for career change. This can be double-edged, however, because a career change now may require relocation later.

In the case of Sylvia and Sam, Sam was willing to relocate but doubted that he could find a situation more to his liking. He understood Sylvia's resentment but at the same time felt that she was less committed to her career than he was and felt it was unfair to make him move. Initially his attitude infuriated Sylvia, but she knew he was right on one key point—she lacked real excitement about her field.

Our discussions about how she had decided on accounting revealed that Sylvia's family situation had been more influential than she had realized. She was the oldest of four children and came from a family with modest income. She wanted to be financially independent as soon as possible and was good in math. Accounting seemed the perfect choice. She did well in her courses and was quite successful in her work. But she knew all along that methodical, routine work bored her; she most enjoyed the interactions with clients and the troubleshooting.

Sam and Sylvia reached a workable, albeit temporary, solution. She met with a career counselor and carefully considered the match between her interests and abilities, on the one hand, and careers in business that could build on her accounting background, on the other. She then enrolled in a master's program in management at a local university. They plan to reevaluate their situation when she finishes.

Differences in the pace at which each spouse pursues life goals can create an imbalance, and in some cases what appears to be to an impossible gulf. The prototypical case involves "reentry women"—that is, women who enter or return to careers at midlife after many years of attending to the home, children, and husband. If these couples shift from a "traditional" single-career family to a "traditional" dual-career family, the transition takes place quite smoothly. The wife continues doing the things at home and does her career as well; the new patterns leave the husband's life altered somewhat but not drastically. But should her career be particularly demanding or should she significantly reevaluate her values and life goals, the transition becomes more complex. The foundation of the marital relationship gets shaken and some kind of basic change may be inescapable.

The case of Jane and Layton illustrates this pattern. Layton, a 55-year-old man in what had been a traditional marriage, explains the situation as follows: "I always knew my wife was very capable and that she would want to work after the children were grown, but I never thought she would get caught up in a demanding career at fifty. I am ready to spend more time together and she is never here."

Layton, a physician, was pleased with his life, until a few years ago. Jane, his wife, was an enthusiastic and loving mother who just about single-handedly raised their three children in addition to keeping an active profile in community volunteer work. They had assumed Jane would return to teaching eventually, and both were pleased when she unexpectedly received an offer to manage a small real estate business. That was three years ago. She has since been promoted twice and now manages a much larger operation, which requires constant travel.

The husband in this example has achieved a personal summit and feels ready to reap the rewards. Layton wants to let up a little and relax more. He comes home for dinner every night, is more leisurely in the morning, and wants his wife to enjoy these times with him. Instead he sees himself facing a situation in which his wife is totally immersed in a demanding career that neither of them had envisioned.

Jane and Layton each have their struggles. She resents fixing his breakfast and dinner and leaving instructions for the microwave before rushing off to a 7 A.M. flight, yet feels bad if she doesn't. Finding a balance between the roles she has always filled and the new roles she is creating causes her considerable conflict. Layton wants to unselfishly support Jane but at the same time feels pushed at midlife into changes in personal values and life goals that he did not consider or invite. His dissatisfaction with their life together brought them to counseling.

Resolution in this case involved Layton's seeing Jane's side of things. Essentially, she said it was her turn. Yes, he resented her not being there but he greatly appreciated her sensitivity to the changes this required in him. Although he never will be the kind of man who has dinner waiting for a tired spouse, he now looks forward to her coming home and takes great interest in her daily activities at work. (Ironically, he takes great delight in his three daughters—all of whom are in role-sharing marriages.) He has rekindled an interest in stamp collecting and gardening, travels with Jane as much as possible, and plans vacations that take advantage of all her frequent-flier miles.

Struggles with Role-Sharing

Oftentimes, spouses wish that they felt differently than they do about who does what in a marriage. They may deny deeply ingrained traditional beliefs and values about male prerogative and female nurturance.

Let's consider the arrangement of John and his spouse, Susan. John was willing to help out with his son and the housework, but his career is more than nine to five. He felt that he is the one with the more demanding and financially rewarding career and that the effort he puts into his career entitles him to be freed from most household and family responsibilities. Susan, on the other hand, saw herself in a no-win situation. Yes, she works fewer hours and makes less income, but is that justification for her doing everything at home?

In this situation, the husband feels deserving of relief from domestic responsibilities. Not unlike many other men, he expects his wife to want to do what's best for him and his career. The wife, not unlike many women in two-career families, has become especially sensitive to husbands' subtle coercive maneuvers to maintain their privileged position. Susan saw John hiding behind his role as a family provider, his attitude being that he could do less family work and that should suit her just fine.

When the couple sought counseling, John felt unsupported in his career and Susan felt unsupported at home. Moreover, Susan questioned staying in the marriage; she felt totally disregarded. John's apparent insensitivity, coupled with his inability to understand her situation, further estranged her from the marriage.

Resolution of their marital conflict required that *both* Susan and John see a way in which their needs could be met. Conflict creates resentment, self-defensiveness, and feelings of being overpowered, disregarded, or defenseless. When a couple is enmeshed in a conflict, only two solutions appear possible—stick to your guns and win or give in and lose. Unfortunately, neither of these solutions provides a resolution.

Moreover, their conflict had to be looked at from *both* their perspectives. Both partners need to know someone understands their position. There is no right or wrong; both partners have legitimate wants and frustrations and neither needs necessarily to "give in" or feel co-opted. Also crucial to reaching a resolution is that partners develop sufficient conviction that they can negotiate for what they want without knuckling under when the going gets rough. Only then can each partner nondefensively respect the other's views and experiences and listen to what he or she has to say.

In this case, Susan needed to know that hearing John out would not result in her automatically relinquishing her view that he was not doing his fair share. She learned from past experiences in the marriage that showing empathy got her nowhere: He often did not reciprocate her support of and sensitivity to his viewpoint. If anything, her sensitivity seemed to undermine her own position. She continued to do everything at home. John, who is very articulate and logical, would convince her that of the two, he was much more overburdened by work. Susan would agree, but not without increasing her resentment.

This couple's feeling that I understood them facilitated discussions between the partners. Each was in a better position to consider the other's perceptions and alter his or her own. Susan understood how important success was to John, but felt that he had to realize his success could not come at her expense; there were many things she was willing to do to support his career, but doing everything at home was not one of them. Her increased openness to him in turn facilitated changes in his rigid stance. John initially said he had felt justified in reneging on role-sharing because he earned more. During counseling, however, it became apparent that his behavior was connected to another resentment in the marriage. He felt Susan also had reneged on the initial marital contract. She was underemployed and lacked the gumption to find a more challenging position. He cared less about her earning more than about her using her abilities to a larger extent. Susan admitted there was some truth in what he said. She in fact had dismissed the possibility of a more challenging position because the domestic responsibilities were on her shoulders.

These discussions freed Susan and John from entrenched positions and allowed them to consider alternatives. Open, honest discussions of what was preferred, needed, and workable became possible. Each felt the other had not lived up to their agreements when they married nine years before and admitted to resistances somewhat rooted in traditional sex roles—his wish to be free of domestic responsibilities and her reluctance to develop her career. They worked out what both considered to be equitable responsibilities at home and Susan looked into and accepted a more suitable position—a success story to be sure, but one not all that unusual for couples who seek counseling because they want their marriage to endure.

Deciding Not to Have a Child

I mentioned earlier that dual-career spouses typically postpone the decision to have a child to their late 20s or early 30s, particularly if the wife gives no signs of wanting to interrupt her career for child-rearing. (Women rarely wait for signs from men to do so.) Delaying gives both partners time to establish themselves in their careers and be in a better economic position to start a family as well as allows both partners and the marital relationship time to mature.

Very few partners seek therapy when they are trying to determine the best time to fit a child into a dual-career family. But the decision to postpone may raise the question of whether to have a child at all. Couples in this quandary often come to counseling to explore the possible ramifications of remaining childless.

Mary and Mike's situation was not that uncommon in that both partners were somewhat indifferent—they could go either way but feared making the wrong decision. Mike is a 39-year-old economist in a role-sharing marriage. Mary, who is a year younger, teaches sociology at a university. Mike described their attitude toward having a child: "When we married we thought we would eventually have a child and every once in a while we give it serious thought. But neither of us gets too passionate about it and we coast along undecided."

This couple felt they needed to make a decision one way or the other. During counseling Mike admitted that he saw the decision as Mary's. He thought having a child would be nice but not essential. He likes kids but never saw himself as centering his life around them. He grew up in "a typical middle-class family—two kids, two cars, and an O.K. neighborhood"—and has few complaints about his childhood. He left home to attend college and never really went back. He sees his parents on occasion but does not feel especially close to them, although he does feel close to his brother.

Mary agreed that the decision mostly rested with her. Coming from a large and close Catholic family, she felt odd about her hesitation and ambivalence. She is the only one of the four siblings without a child, yet she married first. What made the situation so strange in her mind was that she was *not* torn between career and family, nor did she fear giving up one for the other. She was highly visible and successful as a sociologist but not totally consumed by her career. Moreover, she felt confident that Mike would share fully in the child-rearing. That had never been in question. Rather, she wondered if she wanted to raise a child, given all the other concerns she and Mike shared. Their concerns about overpopulation, poverty, and famine headed the list.

Following many months of discussion, they made a conscious decision not to have a child. They left open the possibility of adopting a child—particularly a minority child—although that too seemed very unlikely. Mike looked into getting a vasectomy and had the procedure performed about six weeks later. That was over a year ago. They report much relief and satisfaction. It was a good decision for them.

Accommodating Aged Parents

Because people now live longer, many more couples than formerly are members of the "sandwich generation," middle-aged adults caught between the demands of young adult or teen-age children and aging parents. And the burden of aging parents presently falls more on daughters than on sons or, should there be only sons, on daughters-in-law.

Women nurture more than men and are expected to. Taking care of the aged frequently comes under their purview. The gradual yet constant increase in life expectancy has its pluses and minuses. People with good health can expect additional fine years; those less fortunate can anticipate needing additional care.

Celeste and John began to worry when Celeste's father died suddenly of a heart attack at the age of 72. Her mother was in good health but was quite an unassertive and passive person. Although she had a college degree, she had never worked and had had no desire to. In the five years following his death Celeste's mother made extended visits "to get away from the cold," and this was a quite pleasant arrangement for all concerned. Her mother relished time with the family, loved helping out in any way she could, and looked forward to returning to her own home. The next five years were trying, however. One health problem after another occurred. The children hated coming home after school because grandma was always sick. Celeste and John hesitated to go out of town, which they needed to do frequently for their work, and felt frantic when they were at home. Celeste lost time at work and never knew when she might get a call. Their marriage was taking the brunt of it; tempers flared and communication became strained. Celeste, a businesswoman, explained their situation: "It never occurred to me that I would spend my forties taking care of my mother. As my children became more independent my mother grew increasingly dependent."

Celeste kept hoping her mother would mention moving to a nursing home but she never did. Instead she "visited" for longer and longer periods of time. She never brought up the subject and everyone pretended that grandma was going to get better. John saw the emotional toll all this was taking on Celeste and their family life and suggested they seek professional help.

Counseling helped Celeste and John realize that the decision of what needed to be done was theirs to make, not Celeste's mother's. Her mother assumed someone would take care of her, although she did not provide care for her own parents in their last years. It was not within her personality or life experience to make plans for herself. Thus, the question shifted from "Can we get mom to decide what to do?" to "What kind of care is needed and how can it be best provided?" The solution then became obvious—she would soon need full-time nursing care. After researching the situation, they put her name on the waiting list of an excellent home. The home was just a few miles from where they lived—a stroke of luck that greatly eased the situation.

They broke the news to Celeste's mother gently but firmly and discussed all the possible options. They acknowledged that they both knew she would prefer living with them, but that just was not possible. Emotionally charged as this was for everyone, both Celeste and her mother knew it was the best option. That was three years ago and her mother continues to hang in there.

Hindsight always brings wisdom. Celeste now realizes that she and John should have helped her mother plan for her old age 15 years ago when her father died. Much of their pain could have been avoided had she moved into a retirement home with nursing care facilities when she was still in good health. She would have developed a network of peers and had health care readily available. They now see the importance of planning for their own retirement and care in old age. Also, they openly discuss these matters with John's parents, who, although now in quite good health, may some day need nursing care.

What to Look for in a Therapist

> My husband assumed the female psychologist recommended to me by a feminist friend was a man-hater; I thought the man recommended by his chauvinistic boss was entirely unacceptable. [A 35-year-old woman married ten years]

Spouses seeking personal or marital counseling should first inquire about prospective therapists' credentials, when they had their professional training, what models they use for treatment, and what knowledge and attitudes they have about the two-career family lifestyle. Pertinent questions to ask therapists include:

- What views do they hold about the rights and roles of women and men?
- Whom do they typically see in counseling?
- How often do they work with spouses in dual-career families?
- What theories guide their counseling? How tied are they to tradition?
- What theories guide their views of female and male development? Do they see the sexes as "opposite" for the most part?
- What about their own lifestyles—have they married, divorced, reared children?

What are the proper responses to look for? How do couples determine they have found the "right" therapist? Basically, they must feel respected, understood, and challenged. The therapist must convey an understanding of the conflicts or feelings presented, be free of apparent biases about the roles of women and men, and be knowledgeable about life in two-career families.

Responses to these kinds of queries reveal, for example, possible biases and gaps in professional knowledge, which in the long run could prove detrimental to treatment. Many psychologists lack education and training pertinent to current theories of female development and the changing roles of women and men. Victims of their own socialization, they see the struggles of dual-career couples through a traditional lens and unwittingly may do more damage than good. Therapists opposed to or ambivalent about women pursuing careers, for instance, may interpret problems at home as the woman's fault and encourage a woman seeking help for role conflict to leave her position rather than to explore other aspects of the conflict as well as the husband's role in child care. Similar problems emerge for men who seek close relations with children or couples who struggle with role-sharing should the therapist hold contrary views buttressed by traditional suppositions.

It is especially crucial that the couple feel the therapist understands them. I have seen a number of couples in therapy who expressed relief about seeing someone who did not dispute men's involvement with children and family work and who also understood the "politics of housework." Often the core issue that dual-career couples bring to therapy hinges on desired yet resisted changes in the traditional sex-role division of labor.

A case in point is one distressed couple who could readily decide on who would cook when, but failed week after week to carry out the agreed-upon plan. What emerged in therapy was the husband's irrational rage every time his turn came around. He felt terrible about his resistance and helpless about changing it. Moreover, he was reluctant to share his feelings with his wife because he genuinely believed the plan was fair. He also was loath to admit his traditional views and chose instead to avoid the situation. Admitting to himself and to his wife that he was not yet the liberated man he wished to be freed him to talk about his struggles with household work. She conceded she had no idea how hard it was for him—and this acknowledgment from her made all the difference in the world.

The counselor or therapist must be able to go beyond surface issues and recognize that the experiences of men in the dual-career situation are not the same as those of women. Because the gender-role socialization of women and men differs so dramatically, the areas they experience as problematic, and the factors that contribute to problem areas, often differ markedly. Combining a career and family is a difficult process, involving different choices and changes, for both men and women.

For men, the traditional structure of professional careers and the assumptions associated with male entitlement, superiority, and independence present obstacles to fuller involvement in family life. For women, the traditional division of labor and assumptions about female dependency, inferiority, and nurturance present obstacles to fuller involvement in professional careers. These differences show up in the day-to-day conflicts of men and women in dual-career families and the issues they bring to therapy. Husbands typically struggle with esteem issues stemming from a perceived loss of power and prestige or decreased freedom in their occupational pursuits, competition with their spouse, or participation in "women's work" within the family. Some men, not unlike the husband I just mentioned, discover a fundamental discrepancy between their generally liberal attitudes and their actual behavior within the marriage; it is one thing to support or advocate being a supporter or advocate, but quite another to actually be one. Wives, in contrast, often struggle with esteem issues regarding legitimate expectations for a man's involvement in family work, redefinitions of their roles of wife and mother; and conflicts between parenting responsibilities and career interests.

Finally, the therapist one chooses must be able to look beyond the psychological or personal barriers to achieving a satisfactory dual-career marriage. The therapist must also be cognizant of the external or structural barriers to egalitarian marriages. That men earn more than women on the average, that some women experience prejudice and discrimination in their work, and that child care remains problematic in our society all significantly impact the well-being of dual-career spouses. These men and women face unique problems and encounter predictable obstacles; the counselor or therapist selected must be sensitive to the real difficulties, both psychological and pragmatic, of developing new concepts of maleness and femaleness and of coordinating career and family. Moreover, the therapist needs to respect the values implicit in the choice of this nontraditional lifestyle.

CHAPTER 9

WHAT LIES AHEAD?

What can be said about the future of the two-career lifestyle? Where do we go from here?

The dual-career family is here to stay as a significant alternative to traditional marriage. Increasing numbers of women and men perceive the personal and societal costs associated with traditional views of the sexes, family, and work. In the past, men had to give up emotional expression, intimacy, and relationships with wives and children to attain the power and achievements that defined the traditional male role. Many professional men had no family life; essentially, they sacrificed family life for careers and depended on wives to nurture and sustain them as well as any children. For many, the career man with a loving, close family was largely a myth.

Women were forced into an equally narrow role of nurturance. As described earlier, femininity became associated with dependency on men for love and economic survival. Because the stereotypic woman's place was with her child and in the home, she was denied opportunities thought to be inappropriate for women. Preparation for and devotion to occupational work was one such area. If the marriage failed or the husband left, which was not that infrequent, the woman managed by taking low-paying and relatively dead-end jobs.

Women and men who enter two-career marriages desire a different life for themselves. They envision a more egalitarian society in which changing conceptions of the nature of work and family make the two compatible, not mutually exclusive. Many spouses are striving for more balanced lives and more mutually loving and supportive marriages in which partners see each other as esteemed human beings, not just actors playing roles. Some believe the dual-career family puts more emphasis on family life. When both spouses work, home life takes on more importance and greater salience. It becomes a place for reconnection and revitalization. Yet it also means a sharing of the responsibilities at home. Knowing that one has family responsibilities orients partners toward taking an active role with the children and sets an expectation for attending to and doing for others. One couple mentioned that they both pay a lot of attention to what's happening at home because that's the expectation they've set up for themselves after working all day. When all goes well, they both take credit. And when crises or problems emerge, they both feel responsible.

Women in two-career families want equity in the home as well as in the workplace. Many desire role-sharing in all aspects of the marriage. In response, the 1990s may witness a new wave of changes, one exerted by interested groups of men. Men are developing their own support systems in the form of men's groups, organizations, and alliances that affirm and legitimize their quest for a new male identity.[1] Many men seek these requisites: the right to father with nurturance and expressiveness and to be an intimate member of the family; the courage to share anxieties, sorrows, and joys; and the strength to evolve beyond the traditional male role.[2]

"Sharing it all" does not fit everyone's hopes, needs, and projected goals, however. Not all women and men are geared to dual-career marriages or would wish to choose this option. Many partners, even those in two-career families, follow a more traditional pattern when it comes to home and family. In these families, husbands participate minimally in family work and their preeminence remains unchanged. Still others prefer not to commit themselves to relationships with children.

In addition, society in general remains rigid in its views. It does little to facilitate a lifestyle in which both wife and husband engage themselves in day-to-day family life and at the same time pursue satisfying careers. The changes still necessary will not come without strain and conflict. Although the times in some respects appear favorable to social change, as the prevalence of the two-career family itself attests, current social policy remains susceptible to pressures from conservative elements.

The two-career family, like other manifestations of social change, brings both solutions (such as less rigid sex roles) and problems (such as child care). Moreover, the consequences of social change cannot always be anticipated. (The full effect on all members of the family is yet to be evaluated.) Some people view the two-career family as the key to an egalitarian society, free of oppressive roles based on sex. Others see it as a force destructive to male–female relationships, the family, and society. These opposing views accentuate possible dilemmas between partners of a dual-career marriage and within each one's own mind.

Equity and the Personal Values of Partners

I discussed earlier the importance of equity to marital quality. Equity refers to a feeling of fairness based on partners' perceptions of the overall balance achieved in their relationship. In determining equity, spouses essentially make a comparison between their own cost/benefit ratio and that of their partner. When the ratios appear about the same, partners generally feel satisfied. Dissatisfaction or distress occurs when the ratios seem disparate. Such feelings motivate one or both partners to restore the equity or at the very least reduce the discrepancy. For example, to create a sense of balance, one or both spouses might take on more of the housework, cut back on out-of-town travel, or change work hours. From the viewpoint of equity, then, a sense of fairness rather than a condition of strict equality prevails.

Fairness, of course, lies in the eye of the beholder and therefore remains subject to societal attitudes and practices that assume male dominance and privilege. Take the example of time, a very precious commodity in two-career families. Is an hour of a woman's time worth as much as an hour of a man's time? The answer depends on the criteria used. Money? Men earn more than women on the average. Time available? In the ultimate context, women live longer than men. Motivation? It depends on the task.

There is, of course, no one right, objective answer to each such complex, subjective question. Partners must thus find solutions that fit their particular situation. Danger enters in when partners fail to recognize how outside influences in the form of societal pressures and realities unduly influence or override their personal preferences. Many partners in two-career marriages, for instance, want "to share it all" but find this difficult to achieve, not because of their personal views, but because of existing inequities and resistances in the society at large.

Partners need to be cognizant of how external factors influence their personal choices and well-being. Rather than always blaming themselves or their spouses or feeling inadequate, couples must recognize when stresses stem from insensitive or outmoded social or corporate policies. Take the example I used earlier—

the inability of a two-career couple to provide their child with transportation to afterschool activities. The parents can blame their career ambitions and see this as their personal problem. Alternatively, they can view it as a community problem requiring a community solution, e.g., providing adequate public transportation. The difficulty shifts from a family problem to a societal problem. Because solutions to "external" problems prompt a different set of solutions, this refocusing opens up new possibilities with their own sets of strategies. The parents in this example could join with like-minded parents to educate school boards, employers, or legislators about the need for afterschool transportation or become members of organizations such as the Child Care Action Campaign that have philosophies and goals concordant with those of partners in two-career marriage.

Similarly, dilemmas emerge when there is conflict between a couple's personal values and society's values—as reflected by corporate or government policy. Many partners, for instance, hold egalitarian values about the importance of a career for each spouse. But their view deviates in many cases from the extant social value, which attributes career preeminence to the man. Thus, couples may be unable to act in accord with their personal views because policies such as rules against hiring spouses, higher male salaries, and lack of paternity leave perpetuate hiring policies that favor the husband's career.

Here again, couples need to recognize the role society plays in creating their dilemma. The situation calls for longer-term strategies directed at challenging the assumptions underlying existing policies. For instance, some businesses have appointed task forces to study the effect of specific company policies on dual-career families; others have undertaken self-studies to consider the effects on family life of major organizational policies such as promotion practices.[3] These activities don't come about accidentally; they typically result from pressures applied by employees.

Child Care and Conceptions of Career and Family

Parents will not succeed in giving their children a high priority in two-career marriages without a struggle. Society in many ways pays lip service to the importance of parenting. Changing this situation essentially requires a major rethinking of sex roles within the context of the family and work and within society. The United States is one of the few major Western countries to have no policy for guaranteeing women or men the right to maternity or paternity leave, no job protection in the event of such a leave, no cash benefit to help compensate for not working due to childbirth, and no care for the newborn.[4] Moreover, the idea of modifying what is now considered "professional" behavior gets little serious attention. Traveling, working long hours, and subordinating oneself to one's career remain the norm, and those who follow this prescribed form often get well rewarded. But it's a norm that assumes that someone else, someone other than the parents, carries out family obligations. Such widely held views of careers as these add to the misconception that parents in two-career families abandon children to day-care centers or indentured servants because they themselves become totally absorbed or engulfed by professional work.

In addition, some societal resistance to changes in the roles of women and men gets couched in concerns about child care. I have heard repeatedly the observation that the greater contentment of women comes at the cost of children who need their mothers' care—and not their fathers'. I have also listened to many variations of a comment I overheard not too long ago: "I am tired of hearing about women and their problems. Does no writer on the subject of careers versus motherhood have the temerity to tell a probable truth: that many women complain or quit to raise children purely and simply because working is too tough." Women are now in a Catch-22 situation—seen as either wanting too much or lacking in some way.

Many who associate gender equality with social and family disintegration harbor fears about what male–female equality would mean for them personally. While they loudly vocalize their worry about the children, they often fail to mention their deeper biases involving resistance to changing views of maleness and femaleness. Attaining more egalitarian roles between women and men involves more than sharing status and power. As I argued earlier, spouses in dual-career families also need to revise their sense of self. This is not a one-time task, and it is not easily accomplished, but many of those who engage in it do change and do find the process personally rewarding. Partners who express little fear about male–female equality have confidence in their ability to embrace alternatives to the traditional male and female roles. They are the ones most likely to "share it all" in combining family and career in a richer, more rewarding and fulfilling life.

NOTES

PREFACE

1. Lucia A. Gilbert, *Men in Dual-Career Families: Current Realities and Future Prospects* (Hillsdale, NJ: Erlbaum, 1985).

CHAPTER 2

1. Fox Butterfield, "The Class of '61 Offers Some Dissertations on Itself," *The New York Times* (July 13, 1986), p. 30E.
2. Holly Ornstein, "In Career Goals, Female Valedictorians

Fall Behind," *The New York Times* (November 8, 1987), p. 7.

3. Cathy Guisewite, "Cathy," *The Austin-American Statesman* (June 28, 1987). Copyright 1987 Universal Press Syndicate. Reprinted with permission. All rights reserved.
4. Jeanne H. Block, *Sex Role Identity and Ego Development* (San Francisco: Jossey-Bass, 1984), pp. 253–282.
5. Josephine Humphreys, *Rich in Love* (New York: Viking, 1987).
6. Karen Horney, "The Overvaluation of Love: A Study of a Common Present-Day Feminine Type," *Psychoanalytic Quarterly* 3 (1934), pp. 605–638.
7. Constance Mitchell, "Marriage Rates: Which Study Do You Believe," *The Wall Street Journal* (January 15, 1987), p. 21.
8. Ibid.
9. Colette Dowling, *The Cinderella Complex: Women's Hidden Fear of Independence* (New York: Summit Books, 1981).
10. Harriet E. Lerner, "Female Dependency in Context: Some Theoretical and Technical Considerations," *American Journal of Orthopsychiatry* 53 (1983), pp. 697–705.
11. Alice Adams, "Barcelona," in *Return Trips* (New York: Alfred A. Knopf, 1985), pp. 109–113. Copyright 1985 Alfred A. Knopf. Reprinted with permission.
12. Ibid., p. 43.
13. Julia A. Sherman, "Social Values, Femininity, and the Development of Female Competence," *Journal of Social Issues* 32 (1976), pp. 181–195.
14. "Preliminary Results of the Catalyst Two-Career Survey," *Catalyst Career and Family Bulletin* 2 (May 1981), p. 8.
15. Helen Rogan, "Executive Women Find it Difficult to Balance the Demands of Job, Home," *The Wall Street Journal* (October 30, 1984), pp. 35, 55.

CHAPTER 3

1. D. H. Bell, "About Men," *The New York Times Magazine* (July 31, 1983), p. 46.
2. Gloria Emerson, *Some American Men* (New York: Simon and Schuster, 1985), pp. 230–315.
3. Ibid., p. 315.
4. Anthony Astrachan, *How Men Feel: Their Responses to Women's Demands for Equality and Power* (New York: Anchor Press/Doubleday, 1986).
5. Lucia A. Gilbert, *Men in Dual-Career Families: Current Realities and Future Prospects* (Hillsdale, NJ: Erlbaum, 1985), p. 114.
6. Michael A. Toth, "The Buffalo are Gone: The Decline of the Male Prerogative" (paper presented at the Pacific Sociological Association Annual Meetings, San Diego, CA, April 21–24, 1982).
7. Lawrence Wright, "I Want to Be Alone," *Texas Monthly* (December 1985), p. 166.
8. Joe Kane, "Star Wars: How Men Are Coping with Female Success," *Ms.* (September 1985), p. 52.
9. Arnold Kahn, "The Power War: Male Responses to Power Loss Under Equality," *Psychology of Women Quarterly* 8 (1984), pp. 234–247.
10. J. J. Sempé, *The New Yorker* cartoon (November 8, 1982), p. 117. Copyright 1982 by *The New Yorker.*
11. Diane Baumrind, "New Directions in Socialization Research," *American Psychologist* 35 (1980), pp. 639–652.
12. Letty C. Pogrebin, *Family Politics: Love and Power on an Intimate Frontier* (New York: McGraw-Hill, 1983), p. 197.
13. Diane Baumrind, "New Directions in Socialization Research," *American Psychologist* 35 (1980), pp. 639–652.

14. Joseph H. Pleck, "Men's Power over Women, Other Men, and Society: A Men's Movement Analysis," in *Men in Difficult Times: Masculinity Today and Tomorrow,* ed. R. A. Lewis (New York: Prentice-Hall, 1981), pp. 234–244.

15. Richard Goldstein, "What I Learned about Being A Man," *Ms.* (July 1980), p. 63.

16. Mary Kay Blakely, "He's a Feminist But," *Ms.* (October 1982), p. 44.

17. Rhoda Rapoport and Robert N. Rapoport, "The Next Generation in Dual-Career Family Research," in *Two Paychecks: Life in Dual-Earner Families,* ed. Joan Aldous (Beverly Hills, CA: Sage, 1982), pp. 229–244.

18. Lucia A. Gilbert, *Men in Dual-Career Families: Current Realities and Future Prospects* (Hillsdale, NJ: Erlbaum, 1985), pp. 73–91.

19. Michael Lamb, "Another Look at Nonmaternal Care," *Contemporary Psychology* 29 (1984), pp. 884–885.

20. Lucia A. Gilbert, *Men in Dual-Career Families: Current Realities and Future Prospects* (Hillsdale, NJ: Erlbaum, 1985), p. 75.

CHAPTER 4

1. Ellen Berscheid, "Emotion," in *Close Relationships,* eds. H. H. Kelley, E. Berscheid, A. Christensen, J. H. Harvey, T. L. Huston, G. Levinger, E. McClintock, L. A. Peplau, and D. R. Peterson (New York: Freeman, 1983), pp. 110–168.

2. Robert Sternberg, "The Three Faces of Love," *Psychology Today* (September 1986), pp. 46–54.

3. Margaret Harold, *New Woman Presents Best Cartoons from New Woman* (Fort Worth, TX: New Woman, Inc., 1979).
4. Lucia A. Gilbert, *Men in Dual-Career Families: Current Realities and Future Prospects* (Hillsdale, NJ: Erlbaum, 1985).
5. Philip Blumstein and Pepper Schwartz, *American Couples: Money, Work and Sex* (San Francisco: Jossey-Bass, 1983).
6. Joseph Pleck, *Working Wives/Working Husbands* (Beverly Hills, CA: Sage, 1985).
7. E. Koren, *The New Yorker* (August 20, 1979), p. 93.
8. Jessie Bernard, *The Future of Motherhood* (New York: Dial Press, 1974), p. 162.

CHAPTER 5

1. Margaret M. Poloma, Brian F. Pendleton, and T. Neal Garland, "Reconsidering the Dual-Career Marriage: A Longitudinal Approach," in *Two Paychecks: Life in Dual-Earner Families*, ed. Joan Aldous (Beverly Hills, CA: Sage, 1982), p. 188.
2. Richard Sandza, "A Daughter's Father," *The New York Times Magazine* (February 1, 1987), p. 51.
3. Sharon K. Houseknecht, "Voluntary Childlessness," in *Handbook of Marriage and the Family*, eds. Marvin B. Sussman and Suzanne K. Steinmetz (New York: Plenum, 1987), pp. 369–395.
4. "Pregnancy: Age and Outcome," *Harvard Medical School Health Letter* (October 1985), pp. 1–3.
5. N. Maxine Soloway and Rebecca M. Smith, "Antecedents of Late Birthtiming Decisions of Men and Women in

Dual-Career Marriages," *Family Relations* 36 (1987), pp. 258–262.

6. Carolyn A. Walker, *The Timing of Motherhood* (Lexington, MA: Heath, 1986).
7. Toni Falbo, "Don't Pity the Only Child," *Psychology Today* 9 (1976), p. 60.
8. Garry Trudeau, "Doonesbury," *The Austin-American Statesman* (August 3, 1987).
9. Sheila B. Kamerman, *Parenting in an Unresponsive Society: Managing Work and Family* (New York: Free Press, 1980).
10. "When You Need Child Care Next," pamphlet available from the American Psychological Association, 1200 Seventeenth Street NW, Washington, D.C. 20036.
11. P. K. Knaub, "Growing Up in a Dual-Career Family," *Family Relations: Journal of Applied Family and Child Studies* 35 (1986), pp. 431–437.
12. Lucia A. Gilbert and Sue Lucas, "Predictors of Adolescent Girls' Ego Identity Development" (paper presented at the XIIth International Round Table for the Advancement of Counseling, Lund, Sweden, 1986).
13. Grace K. Baruch and Rosalind C. Barnett, "Father's Participation in Family Work: Effects on Children's Sex Role Attitudes" (Working Paper No. 126, Wellesley College, Center for Research on Women, Wellesley, MA 02181, 1984).
14. Steve Weiner and Bob Davis, "It's Business as Usual at Toy Fair: Familiar Themes and a Harder Sell—Violence for Boys, Love for Girls," *The Wall Street Journal* (February 7, 1985), p. 35.
15. Jeanne H. Block, *Sex Role Identity and Ego Development* (San Francisco: Jossey-Bass, 1984), pp. 253–282.

16. Thomas M. Boyd, "First Love Discovers My Daughter," *The Wall Street Journal* (October 15, 1987), p. 30.

CHAPTER 6

1. Garry Trudeau, "Doonesbury," *The Austin-American Statesman* (January 28, 1987). Copyright 1987 G. B. Trudeau. Reprinted with permission of Universal Press Syndicate. All rights reserved.
2. Lucia A. Gilbert, *Men in Dual-Career Families* (Hillsdale, NJ: Erlbaum, 1985), pp. 22–46.
3. "The Job Market Opens Up for the 68-Cent Woman," *The New York Times* (July 26, 1987), p. 6E.
4. W. Serrin, "Experts Say Job Bias against Women Persists," *The New York Times* (November 25, 1984), pp. 1, 18.
5. Helen Rogan, "Women Executives Feel That Men Both Aid and Hinder Their Careers," *The Wall Street Journal* (October 29, 1984), pp. 31, 44.
6. Joanne Wilkens, *Her Own Business: Success Secrets of Entrepreneurial Women* (New York: McGraw-Hill, 1987).
7. For further information contact: WAGES, Institute for Research on Women's Health, 1616 18th Street NW, #114, Washington, D.C. 20009.
8. Gordon Mott, "Following a Wife's Move," *The New York Times Magazine* (April 14, 1985), p. 58.
9. Lucia A. Gilbert, *Men in Dual-Career Families* (Hillsdale, NJ: Erlbaum, 1985), pp. 105–115.
10. H. M. Azkins, *The New Yorker* cartoon (June 14, 1982), p. 53.

CHAPTER 7

1. Cathy Trost, "Child-care Center at Virginia Firm Boosts Worker Morale and Loyalty," *The Wall Street Journal* (February 12, 1987), p. 25.
2. The Parental and Temporary Medical Leave Act (S. 249) was reintroduced in the United States Senate on January 6, 1987, by Senator Christopher Dodd after failing to receive a Senate hearing in the 99th Congress. A companion measure, the Family and Medical Leave Act of 1987 (H.R. 925), was reintroduced in the United States House of Representatives on February 3, 1987, by cosponsors Patricia Schroeder and William Clay, modified in November 1987, and is up for a final vote in 1988.
3. D. J. Besharov, "The ABCs of Child-Care Politics," *The Wall Street Journal* (March 9, 1988), p. 24.

CHAPTER 8

1. A. Lakein, *How to Get Control of Your Time and Your Life* (New York: Signet, 1974).
2. Susan M. Campbell, *Beyond the Power Struggle: Dealing with Conflict in Love and Work* (San Luis Obispo, CA: Impact, 1984).
3. Robert E. Alberti and Michael L. Emmons, *Your Perfect Right: A Guide to Assertive Living* (San Luis Obispo, CA: Impact, 1984).

CHAPTER 9

1. Lee Shirley, M. H. Hanson, and Frederick W. Bozett, "Fatherhood: A Review and Resources," *Family Relations: Journal of Applied Family and Child Studies* 36 (1987), pp. 337–338.
2. Ibid., pp. 333–340.
3. "Why Should Companies Think about Women" (report available from Catalyst, 14 E. 60th St., New York, NY 10022, 1983).
4. Sheila B. Kamerman and A. J. Kahn, *Child Care, Family Benefits, and Working Parents: A Study in Comparative Policy* (New York: Columbia University Press, 1981).

SELECTED BIBLIOGRAPHY

Aldous, J. (ed.) *Two Paychecks: Life in Dual-Earner Families*. Beverly Hills, CA: Sage, 1982.

Astrachan, A. *How Men Feel: Their Response to Women's Demands for Equality and Power*. New York: Anchor Press/Doubleday, 1986.

Auerbach, S. *Choosing Childcare: A Guide for Parents*. New York: E. P. Dutton, 1981.

Baruch, G., Barnett, R., & Rivers, C. *Lifeprints: New Patterns of Love and Work for Today's Women*. New York: McGraw-Hill, 1983.

Berger, B., & Berger, P. L. *The War over the Family*. New York: Anchor Press/Doubleday, 1983.

Berk, S. F. *The Gender Factory: The Apportionment of Work in American Households*. New York: Plenum, 1985.

Bernard, J. *The Future of Marriage*. New Haven, CT: Yale University Press, 1982.

Blumstein, P., & Schwarz, P. *American Couples: Money, Work, Sex*. New York: William Morrow, 1983.

Burgwyn, D. *Marriage without Children*. New York: Harper and Row, 1981.

Campbell, S. M. *Beyond the Power Struggle*. San Luis Obispo, CA: Impact Publishers, 1984.

Daniels, P., & Weingarten, K. *Sooner or Later: The Timing of Parenthood in Adult Lives*. New York: W. W. Norton, 1982.

Dowrick, S. *Why Children?* New York: Harcourt Brace Jovanovich, 1980.

Falbo, T. (ed.), *The Single-Child Family*. New York: Guilford Press, 1984.

Gerson, K. *Hard Choices: How Women Decide about Work, Career and Motherhood*. Berkeley, CA: University of California Press, 1985.

Gerstel, N., & Gross, H. *Commuter Marriage: A Study of Work and Family*. New York: Guilford Press, 1984.

Gilbert, L. A. *Men in Dual-Career Families: Current Realities and Future Prospects*. Hillsdale, NJ: Erlbaum, 1985.

Hall, F. S., & Hall, D. T. *The Two-Career Couple*. Reading, MA: Addison-Wesley, 1979.

Hertz, R. *More Equal Than Others: Women and Men in Dual-Career Marriages*. Berkeley: University of California Press, 1986.

Holstrom, L. L. *The Two-Career Family*. Cambridge, MA: Schenkman, 1972.

Kamerman, S. B. *Parenting in an Unresponsive Society: Managing Work and Family*. New York: Free Press, 1980.

Kamerman, S. B., Kahn, A. J., & Kingston, P. *Maternity Policies and Working Women.* New York: Columbia University Press, 1983.

Kimball, G. *The 50–50 Marriage.* Boston: Beacon Press, 1983.

Pepitone-Rockwell, F. (ed.) *Dual-Career Couples.* Beverly Hills, CA: Sage, 1980.

Pleck, J. H. *Working Wives/Working Husbands.* Beverly Hills, CA: Sage, 1985.

Rapoport, R., & Rapoport, R. N. *Dual-Career Families Re-examined.* New York: Harper, 1976.

Rice, D. *Dual-Career Marriage: Conflicts and Treatment.* New York: Free Press, 1979.

Roland, A., & Harris, B. (eds.) *Career and Motherhood.* New York: Human Sciences Press, 1979.

Shaevitz, M., & Shaevitz, M. *Making it Together as a Two-Career Couple.* Boston: Houghton Mifflin, 1980.

Veevers, J. E. *Childless by Choice.* Toronto: Butterworth, 1980.

Vogdanoff, P. (ed.) *Work and Family: Changing Roles of Women and Men.* New York: Mayfield, 1984.

INDEX